# THE ROAD
# FROM
# LOS ALAMOS

# Masters of Modern Physics

**Advisory Board**

Dale Corson, Cornell University
Samuel Devons, Columbia University
Sidney Drell, Stanford Linear Accelerator Center
Herman Feshbach, Massachusetts Institute of Technology
Marvin Goldberger, Institute for Advanced Study, Princeton
Wolfgang Panofsky, Stanford Linear Accelerator Center
William Press, Harvard University

**Series Editor**

Robert N. Ubell

**Published Volumes**

*The Road from Los Alamos* by Hans A. Bethe
*The Charm of Physics* by Sheldon L. Glashow
*Citizen Scientist* by Frank von Hippel

# THE ROAD
# FROM
# LOS ALAMOS

## HANS A. BETHE

**AIP**

The American Institute of Physics

。 4219132

PHYSICS

American Institute of Physics
335 East 45th Street
New York, NY 10017-3483

**Library of Congress Cataloging-in-Publication Data**

Bethe, Hans Albrecht, 1906-
  The road from Los Alamos / Hans A. Bethe.
     p.   cm. – (Masters of modern physics)
   Includes index.
   ISBN 0-88318-707-8
   1. Nuclear weapons.   2. Atomic bomb–History.   I. Title.
   II. Series.
   U264.B455   1991
   355.8'25119'09–dc20                                                90-21024
                                                                         CIP

This book is volume two of the Masters of Modern Physics series.

# Contents

## THE FREEZE

## ADVICE AND DISSENT

## NUCLEAR POWER

# FIVE PHYSICISTS

# ASTROPHYSICS

# *About the Series*

**M**asters of Modern Physics introduces the work and thought of some of the most celebrated physicists of our day. These collected essays offer a panoramic tour of the way science works, how it affects our lives, and what it means to those who practice it. Authors report from the horizons of modern research, provide engaging sketches of friends and colleagues, and reflect on the social, economic, and political consequences of the scientific and technical enterprise.

Authors have been selected for their contributions to science and for their keen ability to communicate to the general reader—often with wit, frequently in fine literary style. All have been honored by their peers and most have been prominent in shaping debates in science, technology, and public policy. Some have achieved distinction in social and cultural spheres outside the laboratory.

Many essays are drawn from popular and scientific magazines, newspapers, and journals. Still others—written for the series or drawn from notes for other occasions—appear for the first time. Authors have provided introductions and, where appropriate, annotations. Once selected for inclusion, the essays are carefully edited and updated so that each volume emerges as a finely shaped work.

Masters of Modern Physics is edited by Robert N. Ubell and overseen by an advisory panel of distinguished physicists. Sponsored by the American Institute of Physics, a consortium of major physics societies, the series serves as an authoritative survey of the people and ideas that have shaped twentieth-century science and society.

# *Preface*

This is a collection of my essays, excluding purely scientific papers, written since the end of the Second World War. During this period, the problem foremost in my mind has been—and still is—the nuclear arms race. At the end of the war, many of my colleagues and I who worked on the development of the atomic weapon, felt that we must inform the public about the probable consequences of a nuclear war. We then hoped to try to slow down and, if possible, reverse the nuclear arms race.

I was one of the founding members of the Federation of the American Scientists which campaigned to achieve these goals. I have also worked with the Union of Concerned Scientists on arms control, but I have never become a member because it also opposes the generation of nuclear power for peaceful purposes—a position with which I do not agree.

In 1956, I was invited to join the Science Advisory Committee which, after the Soviets launched Sputnik at the end of the following year, became the President's Science Advisory Committee (PSAC). As members, we were asked to advise the government on military policy as well as civilian science and technology.

Soon after the PSAC was established, I suggested that it investigate the feasibility of coming to an agreement with the Soviet Union to stop nuclear-weapons tests as a first step toward more general nuclear-arms control. President Eisenhower welcomed the proposal and subsequently I was appointed chairman of an interagency committee to study the feasibility of a test ban. The positive recommendation of that committee led to an Experts Conference on a test ban in the summer of 1958 in which I participated, and then to a political conference to devise arrangements under which the ban would take effect. The first conference was unsuccessful. Later, in 1963, a limited test ban prohibiting tests in the atmosphere, under water, and in space was agreed to by the Soviet Union, the United States, and the United Kingdom.

Ever since, I have been more concerned with the arms race in general. That President Reagan and General Secretary Gorbachev both agreed that nuclear armaments are absurdly large comes as a modest result of the arms-control efforts in which I was joined by many others. Much remains to be done before the world can feel safe from a nuclear holocaust.

## The Bomb

This first group of essays is about nuclear weapons. "How Close Is the Danger?" was written together with Frederick Seitz and first appeared in the book *One World Or None*, edited by Dexter Masters and Katharine Way, Latimer House Ltd., London, 1947. General Groves, who had been the head of the Manhattan Project, argued that the Soviets could not produce an atomic bomb for about 20 years. A similar prediction was made by Vannevar Bush, who had been the leader of American scientific work in support of the Allied war effort. Scientists who had worked on the uranium project generally disagreed and estimated that the Soviets could achieve an atomic bomb in about five years. The scientists were shown to be correct. Seitz and I talked about "a determined country," meaning the Soviet Union, but at the time it was indelicate to name that country directly.

"The Hydrogen Bomb" argues against the development of the hydrogen bomb and is the most important essay in the section. It was published in *Scientific American*, April 1950. At the time, there had been a spirited discussion about whether a crash program to create the hydrogen bomb was desirable. Edward Teller, Luis Alvarez, and Ernest Lawrence argued that it was necessary, following the explosion of the first Soviet atomic bomb in August 1949. These scientists found enthusiastic support from the Congressional Joint Committee on Atomic Energy. The General Advisory Committee to the Atomic Energy Commission, chaired by Robert Oppenheimer, wrote an emphatic opinion warning against the crash program. This episode is discussed in detail in Herbert York's book *The Advisors*. The Atomic Energy Commission, by a bare majority, agreed with its General Advisory Committee. But other government agencies, especially the State Department, argued in favor of development. President Truman, at the end of January 1950, ordered the development of the hydrogen bomb. He was clearly motivated by the fear that the Soviets might develop their own hydrogen weapon, possibly before the U.S.

The sensible way to proceed would have been the approach recommended by Vannevar Bush in 1952 — to do all the theoretical and engineering work, but not to test it. Bush suggested that we then announce that the

U.S. was prepared to test the weapon, but that we would refrain from doing so, unless some other country (meaning the Soviet Union, of course) did so. Because it yields such enormous energy, a hydrogen-bomb test could be easily detected and its source accurately identified, and since a hydrogen bomb is a very complicated device, it is most unlikely that a country would put it in its arsenal without first testing it. Bush's voice was not heeded. The Soviets tested their device in August 1953, nine months after ours. Some claimed that the Soviets had tested their version of a hydrogen bomb. But Herbert York argues cogently that this was not the case, and it is my belief, based on available evidence, that the Soviets did not test a true hydrogen bomb until late 1955—three years after our test. The *Scientific American* essay was written before any of this was known.

"Brighter Than a Thousand Suns" first appeared in the *Bulletin of the Atomic Scientists*, December 1958. While this essay about Robert Jungk's book is fairly critical for its time, today I would be even more so. Jungk is much too kind to German scientists. Carl Friedrich von Weizsäcker in his book *Bewusstseinswandel*, in the chapter "Die Atomwaffe," confirms my view of German scientists' motives. Weizsäcker was one of the two princi-pal leaders of the German project. He shows why Bohr could not and did not believe Heisenberg. The essay also expresses my opinion on the degree to which scientists should speak out on political matters.

"Ultimate Catastrophe?" was first published in the *Bulletin of the Atomic Scientists*, June 1976. In the summer of 1942, at a workshop attended by atomic-bomb theorists in Berkeley, Edward Teller had the sud-den idea that atomic weapons might ignite a thermonuclear reaction in the earth's atmosphere. This notion was sufficiently alarming that Oppen-heimer, who was the leader of the workshop, traveled to Chicago to discuss it with A.H. Compton, manager of the Manhattan Project effort to produce a chain reaction. After the war, Compton, in an interview with Pearl Buck, must have mentioned Teller's concern and Buck apparently misunderstood what Compton had reported. In fact, a few days after Teller first proposed his idea in 1942, I proved that it was impossible. As part of their work at Los Alamos, Konopinski and Teller worked out an even more thorough proof. This essay places the argument in a generally more accessible form.

## Arms Control

Ever since 1945, many scientists have been alarmed by the nuclear arms race. When my colleagues and I developed the atomic bomb at Los Alamos, we had hoped that it would ultimately be placed under international control. While, unfortunately, I did not write an essay on the attempt, I gave a number of talks around the country encouraging it. Unfortunately, the Acheson-Lilienthal plan for international control, presented by Bernard Baruch to the United Nations, was rejected by the Soviet Union, and the arms race was on.

The aim of these essays was to encourage slowing down or, if possible, reversing the arms race. Even so, I do support a nuclear deterrent of moderate size, and I believe that the most important aspect of such a strategy is that each side's nuclear weapons be invulnerable. Some of these essays argue against exaggerating the danger of a Soviet nuclear attack. It now appears that the chances for agreement on a reversal of the arms race are much improved, chiefly because the Soviet Union is now led by Mikhail Gorbachev.

"The Case for Ending Nuclear Tests" was published in *The Atlantic Monthly*, August 1960. In 1958, I was deeply involved in the effort to ban nuclear-weapons tests. As noted above, I was then a member of the President's Science Advisory Committee (PSAC), established by President Eisenhower after the Soviets launched their Sputnik satellite in 1957. Earlier, Linus Pauling had led an international campaign to stop weapons tests, primarily to avoid further radioactive contamination of the atmosphere. Less well known then were the writings by David Inglis. I believed that a test ban might be a suitable first step toward stopping the arms race. I convinced PSAC of this approach, and President Eisenhower welcomed the idea. With the president's encouragement, PSAC set up an interagency committee, with representatives of the Department of Defense, the State Department, the Atomic Energy Commission, and PSAC itself. I was chairman. The report of this committee persuaded Eisenhower to propose a Conference of Experts from the U.S., the Soviet Union, Great Britain, and France to discuss whether a test cessation could be monitored adequately.

The Experts Conference, held in the summer of 1958, reported that it was possible to monitor tests under a ban. So political negotiations began in the autumn of that year. Toward the end of the year, however, Edward Teller and Albert Latter raised new difficulties with the detection of underground tests, noting that it might be possible to conceal them in giant underground cavities—objections that placed enormous hurdles in the way of negotiations. The U.S. then sponsored a vigorous seismic underground explosion detection program, which ultimately led to decisive improvements in detection capability. In the 1980s, Sykes and Evernden of Columbia University's Lamont–Doherty Geophysical Observatory offered good evidence that nuclear explosions, down to about one kiloton, can be detected and distinguished from earthquakes.

The negotiations that began in 1958 were essentially terminated when a U.S. high-flying, intelligence-gathering U-2 plane was shot down over the Soviet Union. It was not until 1963 that an agreement stopping nuclear tests in the atmosphere, oceans, and in space—but not underground—was concluded between the U.S. and the Soviet Union. After a brief pause, both countries resumed weapons tests—now all underground—at about the same frequency as before the agreement was signed. This essay was written after the end of negotiations in 1960. Reading it now, I find it unduly pessimistic about detecting clandestine nuclear weapons tests.

"Disarmament and Strategy" appeared in the *Bulletin of the Atomic Scientists*, September 1962. The arguments proposed in this essay are still valid. I still believe that counterforce is absurd. Hardening missile silos remains a good method, but it has lost some of its effectiveness because of the greatly improved accuracy of nuclear weapons. Submarine-based missiles, however, are still fully effective because submarines at sea remain exceedingly difficult to find. This essay defends an invulnerable deterrent. Arms reductions should be accomplished by reducing delivery systems so that a first strike offers no advantage. Part of this essay was written as a sequel to the previous essay in this collection. Soviet tests in 1962 showed that a 1959 test ban would have resulted in a military advantage for the U.S.

"Antiballistic-Missile Systems," coauthored with Richard L. Garwin, was published in *Scientific American*, March 1968. At first, I supported ABM and worked on it for several years. In the course of my work, however, I became convinced of the effectiveness of countermeasures—and penetration aids, especially. I concluded that ABM is likely to lead to an increase of offense weapons by both sides. Defense of a hard point—such as a missile silo—is possible.

"Meaningless Superiority," published in the *Bulletin of the Atomic Scientists*, October 1981, is a strong essay against the nuclear arms race. Henry Kissinger remarks that the number of nuclear weapons in a nation's arsenal is irrelevant at the level existing now. Only invulnerability counts.

"We Are Not Inferior to the Soviets" appears in this collection for the first time. It is based on a talk delivered at a symposium of the Forum on Physics and Society, sponsored by the American Physical Society, April 1982, in Washington. During the early 1980s, the Committee on the Present Danger argued—and the Reagan administration concurred—that the U.S. was suffering from a "window of vulnerability," believing that U.S. land-based missiles might be destroyed by a first-strike Soviet attack. Those who agreed campaigned for more nuclear arms, such as the MX, a missile with ten very accurately guided warheads. This position ignored the fact that the U.S. possessed an invulnerable deterrent in its missile submarines. In the spring of 1983, the Scowcroft Committee, appointed by President Reagan, offered an analysis similar to mine.

"The Five-Year-War Plan," written together with Kurt Gottfried and published in *The New York Times*, June 10, 1982, was a reaction against a planned buildup of all kinds of armaments, nuclear weapons especially. The plan, supported by Secretary of Defense Weinberger, was essentially abandoned in 1988, owing to fiscal reasons.

"Debate: Elusive Security" appeared in *The Washington Post*, February 6, 1983. My essay was written together with Kurt Gottfried. For many years, Senator Malcolm Wallop of Wyoming has been in favor of space-based defense against Soviet missiles. His plan was a forerunner of President Reagan's Star Wars space-defense system. Wallop's call for small, single-warhead ICBMs, however, is a reasonable suggestion.

"Space-Based Ballistic-Missile Defense," coauthored with Richard L. Garwin, Kurt Gottfried, and Henry W. Kendall, was published in *Scientific American*, October 1984. This essay argues against Star Wars, the Strategic Defense Initiative (SDI). The book *The Fallacy of Star Wars*, by the Union of Concerned Scientists, offers more extensive arguments opposing SDI. A much more thorough study was prepared by a panel convened by the American Physical Society, published in the *Reviews of Modern Physics*, July 1987. The APS panel, with full access to SDI data, essentially came to the same conclusion—there is as yet no technical basis to decide in favor of proceeding with SDI.

"The Technological Imperative" first appeared in the *Bulletin of the Atomic Scientists*, August 1985. Essentially, the arms race has been driven by the idea that every technical possibility for a weapon must actually be developed and deployed. Under this imperative, inventions introduced by one side drive the other side to match each new device and system. The technological imperative is an immature idea. In an earlier industrial age, it may have been reasonable, but with mature technology it no longer is so. Consider MIRV: It is now recognized that it was a bad idea, even though it was technically feasible. In the end, it decreased security on both sides.

"Reducing the Risk of Nuclear War," written together with Robert S. McNamara, and published in *The Atlantic Monthly*, July 1985, is a fundamental essay, directed mainly against SDI. Paul Nitze's criteria remain sound, namely, that SDI should be deployed only if it is invulnerable and cost effective at the margin. The essay argues for mutual deterrence at the lowest force levels, permitting a reduction of combined nuclear-weapon inventory from about 50,000 to about 2,000. A first step in this direction has been adopted in principle by the U.S. and the Soviets in their joint

START program, in which strategic forces on both sides are reduced to about half of present levels. The essay also advocates that the Soviets agree to lift some of their secrecy. A promising move in this direction has already been taken by General Secretary Gorbachev. In the summer of 1988, the U.S. and the Soviet Union agreed to incorporate reduction levels in the INF treaty, which eliminated nuclear missiles of a range between 500 and 5,000 kilometers.

"Chop Down Nuclear Arsenals" was first published in the March 1989 issue of *Bulletin of the Atomic Scientists*. An earlier version was presented as the Kistiakowsky Lecture to the American Academy of Arts and Sciences in Cambridge, Massachusetts on November 14, 1988.

## The Freeze

Randall Forsberg proposed in 1982 that the arms race be stopped by freezing the stockpile of nuclear arms at then current levels. Many of us felt that the U.S. and the Soviets possessed sufficient quantities of nuclear arms to destroy the enemy many times over. So a simple treaty might be concluded, with some verification, to freeze nuclear arms. Popular referenda throughout the U.S. successfully supported the Freeze. President Reagan's first response was his Star Wars space-defense system—in part, no doubt, meant to divert public support away from the Freeze movement. Later, obvious popular sentiment in favor of ending the arms race induced President Reagan to enter arms-control talks seriously. From the very beginning, General Secretary Gorbachev was devoted to arms control, and so the INF treaty was signed and a basis was established for a treaty that would ultimately reduce strategic weapons by half. The Freeze was a very important expression of public opinion which, in my view, strongly influenced President Reagan during his final years in the White House.

"The Value of a Freeze" was written together with Franklin A. Long and appeared in *The New York Times*, September 22, 1982.

"Debate: Bethe vs. Teller" was published in the *Los Angeles Times*, October 17, 1982.

"After the Freeze Referendum," coauthored with Franklin A. Long, was published in the *Bulletin of the Atomic Scientists*, February 1983.

## Advice and Dissent

Should scientists refrain from working on dangerous problems? In my view, it would be impossible to have all competent scientists on all sides of a conflict agree to cease working on arms research. Instead, scientists with knowledge about dangerous developments should advise their governments, urging them to refrain from pursuing dangerous courses. One example is the advice given by the General Advisory Committee of the Atomic Energy Commission in 1949 not to develop the hydrogen bomb. In the past, the best way to offer such advice to the U.S. government was through the President's Science Advisory Committee which functioned well under Presidents Eisenhower and Kennedy. Whether or not their governments respond to their advice, scientists have an obligation to speak out publicly when they feel there are dangers ahead in any planned weapons development.

"Science and Morality" is the record of an interview with Donald McDonald, published in the *Bulletin of the Atomic Scientists*, June 1962.

"Back to Science Advisers," written together with John Bardeen, appeared in *The New York Times*, May 17, 1986.

## Nuclear Power

I am a strong supporter of civilian nuclear power, while I am against nuclear weapons. The disaster at the Chernobyl nuclear power plant did not change my opinion. We must all recognize that the use of fossil fuels will lead unavoidably to increased production of carbon dioxide and ultimately to a greenhouse effect on earth. Nuclear power is the only well-developed and proven alternative to fossil fuels.

"The Necessity of Fission Power" appeared in *Scientific American*, January 1976.

"Debate: Nuclear Safety," an exchange between Frank von Hippel and myself, was published in the *Bulletin of the Atomic Scientists*, September 1975.

"Chernobyl" appeared in the *Bulletin of the Atomic Scientists*, December 1986.

## Five Physicists

All but one—the essay on Freeman Dyson—appeared originally as obituaries of close friends.

"J. Robert Oppenheimer" was first published in *Science*, March 3, 1967.

"Freeman Dyson" appeared in *Physics Today*, December 1979.

"Herman W. Hoerlin" was published in *Physics Today*, December 1984. It was written together with Donald M. Kerr and Robert A. Jeffries.

"Paul P. Ewald," published in *Physics Today*, May 1986, was written with H.J. Juretschke, A.F. Moodie, and H.K. Wagenfeld.

"Richard P. Feynman" appeared in *Nature*, April 14, 1988.

## Astrophysics

"Energy Production in Stars," published in *Physics Today*, September 1968, is a record of my lecture, delivered when I received the Nobel Prize in 1967.

"How a Supernova Explodes" describes my present work on the mechanism of supernova explosions. It appeared in *Scientific American*, May 1985, and was written together with Gerald Brown.

Many of the essays in this collection were written with others. For most of these, major credit should go to my coauthors.

# About the Author

Hans Albrecht Bethe is one of the world's most respected scientists. As theorist, scholar, author, advisor, and advocate, he has played a key role in shaping modern physics. He has received world-wide recognition for his broad scientific and technical range since 1938, when he discovered the way the sun and stars generate energy–an achievement for which he was awarded the Nobel Prize in 1967.

Dr. Bethe has also made many original contributions to nuclear and solid-state physics, reactor design and safety, shock-wave theory, rocketry, astrophysics, and radar among numerous other frontiers. Beginning with his first great teacher, Arnold Sommerfeld, Dr. Bethe has collaborated with some of the most eminent physicists of this century–Fermi, Oppenheimer, Teller, Feynman, and Dyson, among other notable figures.

Born in 1906 in Germany, he fled the Nazi regime in 1933, eventually arriving at Cornell, where he continued his long and distinguished career as scientist and teacher. During the Second World War, Dr. Bethe was head of the Manhattan Project's theoretical division at Los Alamos. After the war he resumed his academic career at Cornell, where he is now professor emeritus.

A founder of the Federation of American Scientists after the war, he and his colleagues advocated nuclear arms reductions and, eventually, the end of the arms race. Later, he was invited to join the President's Science Advisory Committee to offer the government counsel on military and civilian science and technology policy. President Eisenhower then appointed Professor Bethe as chair of a test-ban study group. Dr. Bethe has also explored the energy dilemma facing the nation and the world. He has concluded that the continued development of civilian nuclear power is the only wise alternative to fossil fuels.

A member of the U.S. National Academy of Sciences, Dr. Bethe has been awarded the President's Medal of Merit for his work at Los Alamos, the Max Planck Medal–presented by the German Physical Society for his achievements in theoretical physics–and the U.S. Atomic Energy Commission's Enrico Fermi Award for his efforts on behalf of the Manhattan Project and for his contributions to nuclear theory. Eight universities in the U.S. and Europe have presented Dr. Bethe with honorary doctorates.

# THE
# BOMB

# How Close Is the Danger?

## WITH FREDERICK SEITZ

One of the views concerning our policy on the atomic bomb that is frequently expressed is contained in the slogan, "Keep the secret!" This immediately raises the all-important pair of questions: Is there a secret? If so, can we really keep it in the same sense that we could keep secret our landing position on the Normandy beachhead?

The first part of this query can be answered immediately. At the present moment the British and ourselves possess knowledge of certain basic scientific facts and production techniques that are not generally known throughout the world. These facts and techniques relate to the design and construction of machines for producing pure light uranium (U-235) and plutonium and of bombs made from these materials.

Granting this reply to the first question, we recognize at once that the reply to the second part has tremendous significance in determining our foreign policy. If, for instance, it is beyond the range of possibility to keep to ourselves the facts which center about the technology of the bomb for more than a finite time, such as four or five years, we must pay more attention to our foreign policy than to any other factor in our national program. Otherwise we may find ourselves alienated in a hostile world, a world in which the proximity of sudden death on a large scale is greater than it ever was in the primordial jungle that cradled the human tribes. But before we can answer the second question—namely, how long it will take for another nation to obtain the knowledge necessary to make atomic bombs—we must analyze the history of our own development.

The process of fission, which makes the present atomic bomb possible, was discovered in Germany in the winter of 1938–1939 and first became known in this country in January of 1939. This is the starting date of our activity in the field of bomb development that was brought to its practical culmination on July 16, 1945, when the first bomb test succeeded. The six and one-half years from fundamental discovery to final application can be broken, in more or less clean-cut fashion, into three separate periods.

PERIOD I.    The first period extended from January 1939, to January 1942, that is, to about the time when Pearl Harbor was bombed. This might be called the period of groping, during which many problems had to be solved: first, the purely scientific problems of devising experimental techniques for investigating the feasibility of a chain reaction and of carrying out the requisite experimental work; second, the problems which center about the procurement of adequate funds and personnel; and third, the problem of maintaining national interest in the issue when the facts were still very hazy and speculative. This was the period when it was most necessary to have men of genius and determination behind the work. It is interesting to note that this was also the period in which the work was carried on by relatively small groups of men at several universities, especially Columbia, Princeton, and California, led by a few of the most brilliant of our scientists. The period ended when it became theoretically certain that the chain reaction would work.

PERIOD II.    This period extended from January 1942, to about January 1944. Activities expanded and shifted from pure research to the construction of the first chain-reacting unit and to the design of plants to produce fissionable material on a large scale. Contracts were made with industrial companies to build or operate large-scale plants. Pilot plants were actually built and produced moderate amounts—a few grams—of active materials. These plants were also used to gain further knowledge for the design of the large-scale production plants. Three different kinds of plants were developed, one for the production of plutonium and two for the separation of the isotopes of uranium. The triple development was considered necessary because it was not known which of the methods would succeed, and succeed in the shortest time possible. It caused much additional industrial and scientific effort as well as added expense. In this second period, the Los Alamos bomb laboratory was established, and the design of the bomb was begun.

PERIOD III.    The last period extended from January 1944, to the summer of 1945. During this time the large plants for manufacturing bomb materials were completed and set in operation. The development passed from the research and pilot-plant stage to large-scale manufacture. The active material produced was used for experiments to determine the size and other features of the bomb, and the bomb design was completed. Bomb research was handicapped by the fact that again several lines of attack had to be followed, just as in the development of production processes. Finally, in July 1945, the test was made to show the general feasibility of the bomb as well as the soundness of the actual design.

We now reach a point at which it is possible to formulate the problem concisely: How long would it take for foreign countries, other than those involved in the British Commonwealth, to go through each of the three stages described above? The Axis Powers may be left out of consideration. The most important countries are undoubtedly Russia and France, but China or Argentina (or a combination of South American nations) may well be considered too. Also the possibility exists that a highly developed small nation like Sweden or Switzerland might make common cause with a great power of less industrial development.

There is no doubt that many of these countries have as much incentive to learn the facts about the atomic bomb as we ever had. Whether a country is justified in any apprehensions it may have is not the problem; in a world of strong national sovereignties, war preparations are made for potential and often remote emergencies. There can be no doubt that in the absence of international control of the atomic bomb, the Russians will try to develop the bomb in the shortest possible time and will devote a large share of their resources to this end. France has already made public the fact that she is starting an atomic-bomb project, with the Sahara as the prospective testing ground and an initial appropriation far greater than that made in this country during the entire first period (1939–1941).

Granting incentive, we may next ask about availability of the scientific talent that will be needed. The United States and Great Britain undoubtedly contain a lion's share of the outstanding scientific talent of the world at the present time. Moreover, this talent has had in the past six years very good conditions under which to work. These two facts explain why we succeeded in developing the bomb in six and one-half years. It would be difficult to argue that any other nation or combination of nations could have done the job faster *if it had started from the same point as we did in 1939*. On the other hand it would be equally difficult to argue that no other country could have accomplished what we have in any period of time. In the first place, Russia and France both have men of outstanding ability. In the second place, it should be recognized that during periods I and II, in which the major portion of our advance was made, the principal work was in the hands of a small number of people; that is, a large number of good men is not an essential factor. It is almost certain that the reason foreign nations did not proceed very far during the period between 1939 and 1945 (if in fact this is the case!) is because they could not or did not devote full attention to the matter. Russia was fighting for her life with only a fraction of the equipment she needed; France was occupied; Germany, which probably came closest to success, considered the war won in 1941 and 1942, and therefore did not pursue this long-range development with vigor.

What about the availability of materials? The initial work in this country was done on a modest scale with facilities provided by universities. Facilities of this type can be found in all the nations under discussion and many others. Pilot-plant work, corresponding to the developments of period II, requires larger quantities of materials, particularly uranium. This mineral is found in appreciable quantity in St. Joachimethal in Czechoslovakia, and also in Russia, Sweden, and Norway, not to speak of the large deposits in the Belgian Congo. We may safely conclude that any nation engaging in the effort will have little difficulty in procuring enough for pilot-plant work.

Considering the ubiquitous nature of uranium, it is hard to believe that a nation with the surface area of Russia, for example, could find difficulty in locating deposits of sufficient size to go into large-scale production ultimately. This item might be an obstacle to quantity production of bombs in some smaller countries; however, it would not prevent such nations from engaging in the type of pilot-plant research that is designed to lead to successful plants and bombs—and perhaps trading the results for a price. Moreover, if uranium becomes a material of prime importance to the life of a nation, it will become also more precious than gold, and mining of very low-grade ores will become profitable from the national point of view. Gold is extracted commercially from gravel containing as little as .3 part of gold per million; the average uranium content of the earth's crust is believed to be about 20 times as large as this, or 6 parts per million.

With respect to industrial capacity, many of the countries mentioned are far advanced. True, the scale of their production is not so large as ours, but $2 billion is by no means an excessive sum for any of them in comparison with their national income in, let us say, five years. *Moreover, as we shall discuss below, it will probably be far less expensive to repeat our development a second time.*

Some data collected by Lawrence Klein of the University of Chicago show that in Sweden—to take one of the smallest of possible bomb-makers—the average annual gross output of plant and equipment (that is nonconsumption goods) in the period of 1925–1930 was $350 million. Much of this was used for replacement of an old plant and equipment, but when the issue is one of preparation for war, this factor can be readily modified. During our own war-production years, we managed to supply sufficient armaments to the United Nations' armies by not making the customary replacements of our plant and equipment. Sweden, in the act of producing bombs, could also consume capital in order to reach her objective.

An all-out effort on the part of Sweden might call for an average expenditure of $200 million per year for five years. In terms of her 1925–1930 output, this would absorb 57 percent of her capacity to produce plant and equipment and only 10 percent of her gross capacity to produce goods and services of all types. These percentages are very small as compared with those required by the war efforts of the United States, the Soviet Union, Great Britain, Germany, etc. Such a program would be a rather simple order for Sweden if she really wanted atomic bombs.

Many Americans believe that the Russians, although capable of quantity production, are backward in the quality of their industries. The writers would like to point out, as just one item bearing on this point, that the Russians carried out an extensive tank program during the war and produced in quantity a tank that was as good as the best German production and, in the writers' opinion, far better than our own Sherman. This program must have involved an effort comparable to that which we put into the atomic bomb, from the standpoint of both technology and production.

In coming to an estimate of the time required for a foreign nation to produce the atomic bomb, we must compare not only its resources with ours but also its starting point with our starting point. Any nation which begins working on the development of the atomic bomb at the present time starts with far more knowledge than we possessed in 1939. The two major sources of information are, first, knowledge that the bomb works and is sufficiently small to be easily transported by air; and second, the Smyth* report which contains some rather specific data.

Consider first the advantages derived from knowledge that the bomb works. Such knowledge means that much of the groping and speculation that was necessary during period I of the three periods discussed above is unnecessary. Thus, the incentive for working very hard and on a large scale from the start is provided immediately. The greatest effort of period I in our development was devoted to obtaining scientific aid and financial backing, and all this time can now be saved. Moreover, it is no longer necessary to depend upon the vision and judgment of the men of rare genius. Quite average scientists can appreciate the factors involved. Further, it becomes pos-

*Smyth, Henry DeWolf, *Atomic Energy for Military Purposes*, vs. U.S. GPO, 1945.

sible to reduce the total time by starting all three phases of the program at once. It is no longer necessary to await the result of the first period of development before deciding how much effort to risk on the second and third periods.

Consider next the Smyth report, which provides detailed qualitative information on the general direction in which work can be pushed profitably. With respect, for example, to the production of plutonium, the Smyth report states that one can expect to operate a reactor pile with the use of natural uranium and a graphite moderator, even though water, which absorbs neutrons and thereby cuts efficiency, is allowed in the system as a coolant. Moreover, it follows from the report that plutonium produced during the reaction can be separated chemically and in sufficient purity for use in a bomb. The report does not furnish precise information on the dimensions of the pipes and other conduits used in the installation, nor does it describe in detail the methods used for chemical separation. However, men of a far lower order of genius than those who planned the original work could undoubtedly fill in the missing pages as long as they are bolstered with the positive knowledge that the entire program is feasible.

At a dinner table conversation during a recent scientific meeting, a competent physicist who had not been associated with any of the developmental work on the bomb was heard to relate to another physicist his inferences concerning the production of plutonium and the general dimensions of the atomic bomb, all of which he had drawn from reading the Smyth report. The agreement with unpublished facts was remarkable, and it is safe to say that there are at least 25 people in each of the foreign countries under discussion who could infer as much as he did from the Smyth report.

Equally important is the information that any one of three different processes will lead to success—the production of plutonium by the chain reaction, the separation of uranium-235 by the electromagnetic method, and separation by the diffusion method. Any country starting now with this knowledge can determine with relative ease which of these processes is the cheapest and most adapted to its own industrial facilities. That means a very great saving in money, probably reducing necessary expenditures well below $1 billion. It means a substantial reduction in industrial and scientific effort because all work can be concentrated on one line.

What reduction in time results from all the knowledge now available? The greatest reduction will, of course, occur in the first period of development. This period required three years for our own groups, who worked for a large part of this time without much financial support and without the knowledge of ultimate success. Having this support and the information of the Smyth report, it is difficult to imagine that men of the quality of Auger and Joliot in France and Kapitza, Landau, and Frenkel in Russia would require as long as we to cover the same ground.

Two years could easily be sufficient for this period.

Regarding the second phase of the work, we can safely say that there is now no risk in beginning the plans for pilot-plant operation at once. Detailed data for the production of such units may not be available at the moment. However—if, for instance, it is decided to produce plutonium rather than to separate U-235—it is known that uranium and graphite will have to be used in quantity. As a result, work on the preparation of these materials can begin at once. Here again the Smyth report is very helpful because it indicates that a certain rather simple process for the manufacture of uranium metal has been successfully used—a fact which took long to establish in our own development. Similar preparatory work can be done if one of the separation methods is selected as the most suitable process. In either case, a site can be chosen immediately for the ultimate location of the pilot units, and all necessary preparations for servicing the units can be started. Thus, perhaps a year after research in period I has indicated the dimensions to be used in the pilot units, these units can be functioning.

We come next to the question of large-scale manufacture. All the reasoning that has been applied to pilot-plant production can be applied here too. The proper sites for processing and purification of materials such as uranium and graphite can be carried along with the corresponding work for the pilot units. Some delay might be caused by the development of a chemical process to separate plutonium from uranium, because such a process can probably be found only after the pilot plants have produced sufficient material to work with. Moreover, at this stage the high development of industry counts most, and other nations may require more time than we did because their industry is either in quality or in quantity behind ours. Even so, we are probably putting the figure high if we allow two years for this period, which is about twice as long as the time we required. Adding this to the three years estimated for the completion of periods I and II, we conclude that manufacture of plutonium or uranium-235 (or both) can be under way in five years at the outside. It is clearly recognized, of course, that final manufacturing can be carried out only by a nation that has suitable sources of uranium.

Finally, we come to the all-important issue of the design and construction of the bomb. The design can start very early in the program, probably relatively earlier than in our own development. Basic information obtained during the first two periods of development, through the pilot-plant stage, will provide the necessary knowledge about the bomb dimensions and methods that can be used to detonate it. This information should be available in about the fourth year, according to our estimate, so that the theory of the bomb will be clearly understood by the time the manufacturing units are beginning to yield the bomb material. With much of the bomb design done in advance, it is unlikely that there will be any major delay between manufacture of material and production of the finished product. A year is certainly an outside limit. Altogether we have, therefore, a total elapsed time of six years before bombs are available—slightly less than the time needed by us, in spite of the fact that we have added a year to take into account the supposedly lesser industrial development of other countries.

Thus we find that other nations will probably be able to duplicate our development in about the same time that we required. The Smyth report, valuable mainly because it states that certain processes that would naturally occur to other scientists were actually successful, will be a considerable help in their program. The most important fact, however, is that our entire program was successful. Even if this bomb had not been demonstrated, it would soon have become known that three major factories were engaged in our program, of very different appearance and with very different machinery, and that all continued to operate, showing that there must be three different successful processes. Much of the most relevant information in the Smyth report could thus have been deduced from other evidence. The main secret was revealed, and the main incentive supplied, when the first atomic bomb was dropped on Hiroshima.

Many factors can enter in to reduce the required time estimated here. For one thing, we have adopted all along the somewhat provincial viewpoint that the nation engaging in the work will be less effective than we have been, and this viewpoint may be entirely unjustified. Also, it should be kept in mind that work in one or another nation may already be much farther along than external facts would indicate. Finally, it must never be forgotten that men of genius in other countries may devise methods which are much superior to our own and which would greatly reduce the time involved; our previous estimates have been based on the assumption that a foreign nation would simply copy our own pattern of attack.

We are led by quite straightforward reasoning to the conclusion that any one of several determined foreign nations could duplicate our work in a period of about five years. The skeptical or nationalistic individual might at this point decide that such reasoning should have little effect upon our foreign policy, because it is possible that in five years we shall be so far ahead of our present position that it will not matter whether or not a foreign nation has our present knowledge.

There are two very grave objections to this viewpoint. In the first place, it is entirely possible that a foreign nation will actually be ahead of us in five years. In the second place, even if we have more powerful bombs than they, our preferred position will be greatly weakened. For it is an unfortunate fact that present bombs are of sufficient strength, if used effectively and in sufficient quantity, to paralyze our highly centralized industrial structure in the space of a single day. Any store of more powerful bombs in our arsenals would be of little value unless we could use them to prevent attack, and this seems to be a very remote possibility. The existence of such bombs might have an inhibiting effect in the sense that the enemy would fear reprisals. However, if history provides any lesson, it is that fear of reprisal has never prevented a war in which the chances for quick victory are as great as they would be if an adversary decided to strike rapidly and in full strength with atomic bombs.

# The Hydrogen Bomb

Everybody who talks about atomic energy knows Albert Einstein's equation $E = Mc^2$: *viz.*, the energy release in a nuclear reaction can be calculated from the decrease in mass. In the fission of the uranium nucleus, one-tenth of 1 percent of the mass is converted into energy; in the fusion of four hydrogen nuclei to form helium, seven-tenths of 1 percent is so converted. When these statements are made in newspaper reports, it is usually implied that there ought to be some way in which the mass of a nucleus could be converted into energy, and that we are merely waiting for technical developments to make this practical. Needless to say, this is wrong. Physics is sufficiently far developed to state that there will never be a way to make a proton or a neutron or any other nucleus simply disappear and convert its entire mass into energy. It is true that there are processes by which various smaller particles—positive and negative electrons and mesons—are annihilated, but all these phenomena involve at least one particle which does not normally occur in nature and therefore must first be created, and this creation process consumes as much energy as is afterwards liberated.

All the nuclear processes from which energy can be obtained involve the rearrangement of protons and neutrons in nuclei, the protons and neutrons themselves remaining intact. Hundreds of experimental investigations through the last 30 years have taught us how much energy can be liberated in each transformation, whether by the fission of heavy nuclei or the fusion of light ones. In the case of fusion, only the combination of the very lightest nuclei can release very large amounts of energy. When four hydrogen nuclei fuse to form helium, .7 percent of the mass is transformed into energy. But if four helium nuclei were fused into oxygen, the mass would decrease by only .1 percent; and the fusion of two silicon atoms, if it ever could occur, would release less than .02 percent of the mass. Thus there is no prospect of using elements of medium atomic weight for the release of nuclear energy, even in theory.

The main problem in the release of nuclear energy in those cases that we can consider seriously is not the amount of energy released—this is always large enough—but whether there is a mechanism by which the release can take place at a sufficient rate. This consideration is almost invariably ignored by science reporters, who seem to be incurably fascinated by $E = Mc^2$. In fusion the rate of reaction is governed by entirely different factors from those in fission. Fission takes place when a nucleus of uranium or plutonium captures a neutron. Because the neutron has no electric charge and is not repelled by the nucleus, temperature has no important influence on the fission reaction; no matter how slow the neutron, it can enter a uranium nucleus and cause fission. In fusion reactions, on the other hand, two nuclei, both with positive electric charges, must come into contact. To overcome their strong mutual electrical repulsion, the nuclei must move at each other with great speed. Louis N. Ridenour in the March 1950 issue of *Scientific American* explained how this is achieved in the laboratory by giving very high velocities to a few nuclei. This method is very inefficient because it is highly unlikely that one of the fast projectiles will hit a target nucleus before it is slowed down by the many collisions with the electrons also present in the atoms of the target. Therefore the energy released by nuclear reactions in these laboratory experiments is always much less than the energy invested in accelerating the particles.

The only known way that energy can be extracted from light nuclei by fusion is by thermonuclear reactions, *i.e.*, those which proceed at exceedingly high temperatures. The prime example of such reactions occurs in the interior of stars, where temperatures are of the order of 20 million degrees centigrade. At this temperature the average energy of an atom is still only 1,700 electron volts—much less than the energies given to nuclear particles in "atom smashers." But all the particles present—nuclei and electrons—have high kinetic energy, so they are not slowed down by colliding with one another. They will keep their high speeds. Nevertheless, in spite of the high temperature, the nuclear reactions in stars proceed at an extremely slow rate; only 1 percent of the hydrogen in the sun is transformed into helium in a billion years. Indeed, it would be catastrophic for the star if the reaction went much faster.

The temperature at the center of a star is kept high and very nearly constant by an interplay of a number of physical forces. The radiation produced by nuclear reactions in the interior can escape from the star only with great difficulty. It proceeds to the surface not in a straight line but by a complicated, zigzag route, since it is constantly absorbed by atoms and re-emitted in new directions. It is this slow escape of radiation that maintains the high interior temperature, which in turn maintains the thermonuclear reactions. Only a star large enough to hold its radiations for a long time can produce significant amounts of energy. The sun's radiation, for example, takes about 10 million years to escape. A star weighing one-tenth as much as the sun would produce so little energy that it would not be visible, and the largest planet, Jupiter, is already so small that it could not maintain nuclear reactions at all. This rules out the possibility that the earth's atmosphere, or the ocean, or the earth's crust, could be set "on fire" by a hydrogen superbomb and the earth thus be converted into a star. Because of the small mass of the bomb, it would heat only a small volume of the earth or its atmosphere, and even if nuclear reactions were started, radiation would carry away the nuclear energy much faster than it developed, and the temperature would drop rapidly so that the nuclear reaction would soon stop.

If thermonuclear reactions are to be initiated on earth, one must take into consideration that any nuclear energy released will be carried away rapidly by radiation, so that it will not be possible to keep the temperature high for a long time. Therefore, if the reaction is to proceed at all, it must proceed very quickly. Reaction times of billions of years, like those in the sun, would never lead to an appreciable energy release; we must think rather in terms of millionths of a second. On the other hand, on earth we have a choice of materials: whereas the stellar reactions can use only the elements that happen to be abundant in stars, notably ordinary hydrogen, we can choose any elements we like for our thermonuclear reactions. We shall obviously choose those with the highest reaction rates.

The reaction rate depends first of all, and extremely sensitively, on the product of the charges of the reacting nuclei; the smaller this product, the higher the reaction rate. The highest rates will therefore be obtainable from a reaction between two hydrogen nuclei, because hydrogen has the smallest possible charge—one unit. (The principal reactions in stars are between carbon, of charge six, and hydrogen.) We can choose any of the three hydrogen isotopes, of atomic weight one (proton), two (deuteron), or three (triton). These isotopes undergo different types of nuclear reactions, and the reactions occur at different rates.

The fusion of two protons is called the proton-proton reaction. It has long been known that this reaction is exceedingly slow. As Robert E. Marshak stated in his article, "The Energy of Stars," in the January 1950 issue of *Scientific American*, the proton-proton reaction takes 10 billion years to occur at the center of the sun. Ridenour pointed out that the situation is quite different for the reactions using only the heavy isotopes of hydrogen: the deuteron and triton. A number of reported measurements by nuclear physicists have shown that the reaction rates for this type of fusion are high.

A further variable governing the rate of the reaction is the density of the material. The more atoms there are per unit volume, the higher the probability for nuclear collisions.

It is also well known, as Ridenour noted, that the reactions would require enormous temperatures. Whether the temperature necessary to heat heavy hydrogen sufficiently to start a thermonuclear reaction can be achieved on the earth is a major problem in the development of the H-bomb. To find a practical way of initiating H-bombs will require much research and considerable time.

What would be the effects of a hydrogen bomb? Ridenour pointed out that its power would be limited only by the amount of heavy hydrogen that could be carried in the bomb. A bomb carried by a submarine, for instance, could be much more powerful than one carried by a plane. Let us assume an H-bomb releasing 1,000 times as much energy as the Hiroshima bomb. The radius of destruction by blast from a bomb increases as the cube root of the increase in the bomb's power. At Hiroshima the radius of severe destruction was one mile. So an H-bomb would cause almost complete destruction of buildings up to a radius of 10 miles. By the blast effect alone a single bomb could obliterate almost all of Greater New York or Moscow or London or any of the largest cities of the world. But this is not all; we must also consider the heat effects. About 30 percent of the casualties in Hiroshima were caused by flash burns due to the intense burst of heat radiation from the bomb. Fatal burns were frequent up to distances of 4,000 to 5,000 feet. The radius of heat radiation increases with power at a higher rate than that of blast, namely by the square root of the power instead of the cube root. Thus the H-bomb would widen the range of fatal heat by a factor of 30; it would burn people to death over a radius of up to 20 miles or more. It is too easy to put down or read numbers without understanding them; one must visualize what it would mean if, for instance, Chicago with all its suburbs and most of their inhabitants were wiped out in a single flash.

In addition to blast and heat radiation there are nuclear radiations. Some of these are instantaneous; they are emitted by the exploding bomb itself and may be absorbed by the bodies of persons in the bombed area. Others are delayed; these come from the radioactive nuclei formed as a consequence of the nuclear explosion, and they may be confined to the explosion area or widely dispersed. The bombs, both A and H, emit gamma rays and neutrons while they explode. Either of these radiations can enter the body and cause death or radiation sickness. It is likely, however, that most of the people who would get a lethal dose of radiation from the H-bomb would be killed in any case by flash burn or by collapsing or burning buildings.

There would also be persistent radioactivity. This is of two kinds: the fission products formed in the bomb itself, and the radioactive atoms formed in the environment by the neutrons emitted from the bomb. Since the H-bomb must be triggered by an A-bomb, it will produce at least as many fission products as an A-bomb alone. The neutrons produced by fusion reactions may greatly increase the radioactive effect. They would be absorbed by the bomb case, by rocks and other material on the ground, and by the air. The bomb case could be so designed that it would become highly radioactive when disintegrated by the explosion. These radioactive atoms would then be carried by the wind over a large area of the bombed country. The radioactive nuclei formed on the ground would contaminate the center of the bombed area for some time, but probably not for very long because the constituents of soil and buildings do not form many long-lived radioactive nuclei by neutron capture.

Neutrons released in the air are finally captured by nitrogen nuclei, which are thereby transformed into radioactive carbon-14. This isotope, however, has a long half-life—5,000 years—and therefore its radioactivity is relatively weak. Consequently, even if many bombs were exploded, it is not likely that the carbon-14 would become dangerous.

The decision to proceed with the development of hydrogen bombs has been made. I believe that this decision settles only one question and raises a hundred in its place. What will the bomb do to our strategic position? Will it restore to us the superiority in armament that we possessed before the Russians obtained the A-bomb? Will it improve our chances of winning the next war if one should come? Will it diminish the likelihood that we should see our cities destroyed in that war? Will it serve to avert or postpone war itself? How will the world look after a war fought with hydrogen bombs?

I believe the most important question is the moral one: Can we who have always insisted on morality and human decency between nations as well as inside our own country, introduce this weapon of total annihilation into the world? The usual argument, heard in the frantic week before the president's decision and frequently since, is that we are fighting against a country which denies all the human values we cherish, and that any weapon, however terrible, must be used to prevent that country and its creed from dominating the world. It is argued that it would be better for us to lose our lives than our liberty, and with this view I personally agree. But I believe this is not the choice facing us here; I believe that in a war fought with hydrogen bombs we would lose not only many lives but all our liberties and human values as well.

Whoever wishes to use the hydrogen bomb in our conflict with the U.S.S.R., either as a threat or in actual warfare, is adhering to the old fallacy that the ends justify the means. The fallacy is the more obvious because our conflict with the U.S.S.R. is mainly about means. It is the means that the U.S.S.R. is using, both in dealing with her own citizens and with other nations, that we abhor; we have little quarrel with the professed aim of providing a decent standard of living for all. We would invalidate our cause if we were to use in our fight means that can only be termed mass slaughter.

We believe in personal liberty and human dignity, the value and importance of the individual, sincerity and openness in the dealings between men and between nations, prosperity for all and peace based on mutual trust. All this is in great contrast to the methods which the Soviet government uses in pursuing its aims and which it believes necessary in the "beginning phase" of Communism—which by now has lasted 33 years. Regimentation of the private lives of all citizens, systematic education in spying upon one's friends, ruthless shifting of populations regardless of their personal ties and preferences, inhuman treatment of prisoners in labor camps, suppression of free speech, falsification of history in dealing both with their own citizens and with other nations, violation of promises and treaties, and the distorted interpretations offered in excuse of these violations—these are some of the methods of the U.S.S.R. which are hateful to the people of the Western world. But if we wish to fight against these methods, *our* methods must be clean.

We believe in peace based on mutual trust. Shall we achieve it by using hydrogen bombs? Shall we convince the Russians of the value of the individual by killing millions of them? If we fight a war and win it with H-bombs, what history will remember is not the ideals we were fighting for but the methods we used to accomplish them. These methods will be compared to the warfare of Genghis Khan, who ruthlessly killed every last inhabitant of Persia.

What would an all-out war fought with hydrogen bombs mean? It would mean the obliteration of all large cities and probably of many smaller ones, and the killing of most of their inhabitants. After such a war, nothing that resembled present civilization would remain. The fight for mere survival would dominate everything. The destruction of the cities might set technology back a hundred years or more. In a generation even the knowledge of technology and science might disappear, because there would be no opportunity to practice them. Indeed it is likely that technology and science, having brought such utter misery upon man, would be suspected as works of the devil, and that a new Dark Age would begin on earth.

We know what physical destruction does to the moral values of a people. We have seen how many Germans, already demoralized by the Nazis, lost all sense of morality when during and after the war the bare necessities of life, food, clothing, and shelter were lacking. Democracy and human decency were empty words; there was no reserve strength left for such luxuries. If we have learned any lesson from the aftermath of World War II, it is that physical destruction brings moral destruction.

We have also learned that prosperity is the best shield against communism and dictatorship, and in this knowledge we have poured billions of dollars into Western Europe to restore her economy. This generosity has won us more friends than anything else we have done. But after the next war, if it were fought with atomic and hydrogen bombs, our own country would be as grievously destroyed as Europe and the U.S.S.R., and we could no longer afford such generosity. It would be everyone for himself, and everyone against the other.

It is ironical that the U.S. of all countries should lead in developing such methods of warfare. The military methods adopted by this nation at the outset of the Second World War had the aim of conserving lives as much as possible. Determined not to repeat the slaughter of the First World War, during which hundreds of thousands of soldiers were sacrificed in fruitless frontal attacks, the U.S. high command substituted war by machines for war by unprotected men. But the hydrogen bomb carries mechanical warfare to ultimate absurdity in defeating its own aim. Instead of saving lives, it takes many more lives; in place of one soldier who would die in battle, it kills a hundred noncombatant civilians. Surely it is time for us to reconsider what our real intentions are.

One may well ask: Why advance such arguments with reference to the H-bomb and not atomic bombs in general? Is an atomic bomb moral and a hydrogen bomb immoral, and if so, where is the dividing line? I believe there was a deep feeling in this country right after the war that the use of atomic bombs in Japan had been a mistake, and that these bombs should be eliminated from national armaments. This feeling, indeed, was one of the prime reasons for President Truman's offer of international control in 1945. We know that the negotiations for control have not led to success as yet. But our inability to eliminate atomic bombs is no reason to introduce a bomb which is a thousand times worse.

When atomic bombs were first introduced, there was a general feeling that they represented something new, that the thousandfold increase of destructive power from blockbuster to atom bomb required and made possible a new approach. The step from atomic to hydrogen bombs is just as great again, so we have again an equally strong reason to seek a new approach. We have to think how we can save humanity from this ultimate disaster. And we must break the habit, which seems to have taken hold of this nation, of considering every weapon as just another piece of machinery and a fair means to win our struggle with the U.S.S.R.

I have reviewed the moral issues that should deter us from using hydrogen bombs, even if we were sure that we alone would have them, and that they would contribute to our victory. As Ridenour explained, the situation is rather the opposite. We can hardly expect to have a monopoly on hydrogen bombs. If we ever had any illusions about this, the events of the past few months should have destroyed them. The U.S.S.R. has the atomic bomb. She was undoubtedly helped in her efforts by the secret information she received from Klaus Fuchs, which presumably included many of the vital "secrets" of our project. But knowing how a group of scientists put the bomb together would not by itself enable a nation to make one. If Fuchs had given his information to Spain, for instance, it would hardly have been understood; it would presumably not have been used, and even if used it would almost certainly not have led to success. The prime requirements for the job still are a group of highly capable scientists, a country determined to make the weapon, and a great industrial effort. We know now, if we ever doubted it, that the U.S.S.R. has all of these. For the Soviet scientists the information must simply have resolved many doubts as to which steps to take next and saved a number of costly and futile parallel developments.

Their obvious competence will presumably again bring success to the Russians when they try to develop the H-bomb. Yet their decisions and their successes are not independent of our own. Our decision to make the H-bomb, which showed that we considered the project feasible, may well have prompted them to take the same decision. For this reason I think that our decision, if taken at all, should have been taken in secret. This became impossible, however, when the advocates of the H-bomb used public statements as a means of exerting pressure on the president. If the Russians were already working on the H-bomb before our decision, they will now have increased their effort.

It is impossible to predict whether we or the Russians will have the hydrogen bomb first. We like to assume that we shall. If so, I refuse to believe that the U.S. would start a preventive war. That would violate all the fundamental beliefs of this nation, and that these beliefs are still strong is shown by the history of the past four years: although we had a monopoly on the atomic bomb we did not start a war. Clearly, then, the time will come when both the U.S.S.R. and this country will have H-bombs. Then this country will be much more vulnerable than the U.S.S.R.: as Ridenour explained, we have many more large cities that would be inviting targets, and many of these lie near the coast so that they could be reached by submarine and perhaps a relatively short-range rocket. I think it is therefore correct to say that the existence of the hydrogen bomb will give us military weakness rather than strength.

But, say the advocates of the bomb, what if the Russians obtain the H-bomb first? If the Russians have the bomb, Harold Urey argued in a speech just before the president's decision, they may confront us with an ultimatum to surrender. I do not believe we would accept such an ultimatum even if we did not have the H-bomb, or that we would need to. I doubt that the hydrogen bomb, dreadful as it would be, could win a war in one stroke. Though it might devastate our cities and cripple our ability to conduct a long war with all modern weapons, it would not seriously affect our power for immediate retaliation. Our atomic bombs, whether "old style" or hydrogen, and our planes would presumably be so distributed that they could not all be wiped out at the same time; they would still be ready to take off and reduce the country of the aggressor to at least the same state as our own. Thus the large bomb would bring untold destruction but no decision. I believe that "old-fashioned" A-bombs would be sufficient to even the score in case of an initial Soviet attack with H-bombs on this country. In fact, because of the greater number available, A-bombs may well be more effective in destroying legitimate military targets, including production centers. H-bombs, after all, would be useful only against the largest targets, of which there are very few in the U.S.S.R.

So we come finally to one reason, and only one, that can justify our building the H-bomb: Namely, to deter the Russians from using it against us, if only for fear of our retaliation. Our possession of the bomb might possibly put us in a better position if the U.S.S.R. should present us with an ultimatum based on their possession of it. In other words, the one purpose of our development of the bomb would be to prevent its use, not to use it.

If this is our reason, we can contribute much to the peace of the world by stating this reason openly. This could be done in a declaration, either by Congress or by the president, that the U.S. will never be the first to use the hydrogen bomb, that we would employ the weapon only if it were used against us or one of our allies. A pledge of this kind was proposed in a press statement by 12 physicists, including myself, on February 4, 1950. It still appears to me as a practical step toward relief of the international tension, and toward freedom from fear for the world. The pledge would indicate our desire to avoid needless destruction; it would reduce the likelihood of the use of the hydrogen bomb in the case of war, and it would largely eliminate the danger that fear of the H-bomb itself would precipitate a war.

If we do not make this pledge, the hydrogen bomb would almost surely be used. Once war broke out, our military leaders would be blamed, in the absence of a pledge, if they did not immediately initiate a full-scale hydrogen-bomb attack. But if such a pledge existed, they would be blamed if they did use the bomb first. To be sure, the pledge might not be relied on by our adversaries, but at least it would create a doubt in their minds and they might decide to wait and see. Perhaps they would not wish to provoke the certain use of the bomb by dropping the first one. Moreover, if they started a war, they would probably hope to capture our country and to exploit its wealth rather than to conquer a heap of rubble.

We have proposed unilateral action rather than an international treaty on this pledge. We have done this because negotiations with the U.S.S.R. are known to be long and frustrating. A unilateral pledge involving only this country could be made quickly, and it could not again lead to the disappointment of a breakdown of negotiations. On the other hand, we certainly would not want to exclude a pact with the U.S.S.R. on this subject. This might be the first point on which the two countries could agree, and this in itself would be important.

Obviously the pledge can only be a first step. What we really want is a workable agreement on atomic energy, as part of our efforts toward a lasting peace. Much has been said about new negotiations on atomic control. Opinions vary from that of Senator Brien McMahon, who proposed to spend $50 billion for rehabilitation of war-devastated countries, including the U.S.S.R., in exchange for an atomic settlement, to that of Senator Millard Tydings, who declared that an atomic settlement would not be acceptable to this country unless it was coupled with general disarmament, which he has advocated for a long time. Both of these viewpoints, and those of many other senators, show the desire of this country for some agreement. At the same time there are persistent reports, clearly indicated in recent dispatches from The New York Times correspondent in Moscow, that the Russians might like to negotiate. It seems to me that too much is at stake to miss any such opportunity.

On the other hand, President Truman voiced the fears of many of us when he stated recently that there is no security in agreements with the Russians because they break them at will. He referred to the agreements of

Yalta and Potsdam in 1945. Since then we have learned much about Soviet methods, and the Russians have found that we do not retreat as easily as they apparently imagined in 1945. This more realistic mutual appraisal makes it much more likely that we could now come to arrangements which neither side would regret afterward. Obviously in any negotiation each side must be willing to make concessions and to consider primarily proposals directed to mutual advantage rather than superiority over the other.

The situation in atomic energy has changed, both because of the Soviet development of the A-bomb and because of our decision on the H-bomb. To leave atomic weapons uncontrolled would be against the best interests of both countries. If we can negotiate seriously with the U.S.S.R., the scope of the negotiations should probably be as broad as possible. But the situation would be greatly eased even if we could agree only to eliminate the greatest menace to civilization, the hydrogen bomb.

# Brighter Than a Thousand Suns

I n the "golden times" of the 1920s and 1930s a handful of physicists came upon new discoveries almost every day. Quantum mechanics was created in those years, and a small number of devoted scientists found how one after the other of the great puzzles of atomic physics and of chemistry could be explained and fell into place. The atomic nucleus was explored experimentally, and here quantum mechanics was again successful: it could predict the rate of nuclear reactions. Chadwick discovered the neutron, and this made it possible to develop a rational theory of the structure of the nucleus.

All these discoveries were made with small, rather inexpensive apparatus. The physicists in all countries knew each other well and were friends. And the life at the centers of the development of quantum theory, Copenhagen and Göttingen, was idyllic and leisurely, in spite of the enormous amount of work accomplished.

How it all has changed! There are now enormous accelerators, with large groups of scientists working on each; a wealth of detailed material is published in highly specialized journals every week so that it has become impossible to keep up with the literature even in a narrow part of nuclear physics; announcements of important discoveries appear first in *The New York Times*. We fly many times every year across the country or across the Atlantic to hold mammoth conferences in which it is difficult to find our friends. The life of physicists has changed completely, even of those who are not involved in politics or in technological projects like atomic energy. The pace is hectic. Yet the progress of fundamental discovery is no faster, and perhaps slower, than in the thirties.

The basis for this fundamental change was laid in the early thirties when particle accelerators were beginning to appear—the Cockcroft-Walton, the Van de Graaff, and the cyclotron. With the further development of these machines, physics would have changed even without any political events.

Robert Jungk, in his book, *Brighter Than a Thousand Suns* (Harcourt,

Brace, 1958) stresses that all the great discoveries of the 1920s and 1930s were made purely for the sake of knowledge, without any thought of technological, let alone military, application. It would be a great accomplishment if his book would contribute to an understanding of this fact by the public, and to an appreciation of the great spiritual adventure of science, which is too often considered the handmaiden of technology.

Jungk's main theme is the impact of politics on scientists and conversely, of science and scientists on politics. Jungk quotes Langevin's words in 1933 that "the neutron will be remembered long after Hitler" and we cannot but agree with the prophecy. He tells about Hitler and his persecution of the scientists of Jewish descent, many of whom later played a decisive part in the development of atomic weapons in the U.S. He tells of the distortion of science itself by the Nazis, and of their anti-intellectualism. And he tells how slowly and painfully, through a comedy of errors, the fission of uranium was discovered, and how this discovery caused a chain reaction among the atomic physicists who immediately started to project its applications.

On the whole, Jungk faithfully reports the history of the atomic bomb. Politically, Szilard appears as a hero: The prime mover in getting President Roosevelt interested in the project, and in imposing voluntary secrecy on scientific publications on nuclear fission, and again the most persistent driving force behind the Franck Report warning against the use of the bomb against Japan, and behind the successful movement to put atomic energy under civilian control in the U.S. Jungk points out how the most compelling motive for the scientists on the uranium project was the fear that Hitler's Germany might succeed in getting the bomb before the West, a motive particularly strong among those scientists who had seen Nazi terror firsthand.

Missing is a description of the flavor of the technical work itself, of the tremendous sense of urgency, the enormous complexity of the manifold tasks, whether we would succeed in solving *all* these problems on time, the sense of accomplishment at each major stage. The history of this fascinating experience is yet to be written. Jungk provides only a glimpse of it in the account of the "Trinity Test" in the New Mexico desert.

A more serious criticism concerns Jungk's description of the German uranium project. I agree with Jungk that the best German physicists on the project, especially Heisenberg and von Weizsäcker, did not have the same sense of urgency as their American colleagues, and had at least a divided mind whether and to what extent they should help their government. As far as I know, however, and contrary to Jungk's presentation, ethical reasons played only a minor role in their decision not to push the development of the atomic bomb. The decision was mainly based on what they believed to be a realistic appraisal of the difficulties: Heisenberg told me that he had estimated that it would be far beyond the capacity of Germany and, likewise of the U.S., to develop an atomic bomb during the Second World War, even though he apparently realized its theoretical possibility. Had the project seemed "technically sweet," German scientists might have had a very different attitude. As it was, they regarded the uranium project mostly as a device to save their colleagues from being killed in the war, an aim in which they were successful. So work went ahead on the uranium project, as fast as the moderate government support permitted. German physicists even believed themselves to be ahead of the Western Allies at the end of the war, and thought their very incomplete heavy-water reactor might be a great asset in the armistice.

Jungk tells how Heisenberg tried to convey to Niels Bohr that the Germans were not trying to develop a bomb but merely a pile. He argues that we should have believed it in the name of international science. But it clearly would have been irresponsible of Western physicists to do so.

Like many others, Jungk argues strongly that the bomb should not have been used against Japan, and puts some blame on the Technical Advisory Committee, consisting of A. H. Compton, Fermi, Lawrence, and Oppenheimer for agreeing to its use, and praises the Franck group for its appeal not to use it. Praise certainly belongs to them but their report—even though quite prophetic—does not, on re-reading today, appear forceful enough to persuade a Jungk government to refrain from using a weapon which was likely to stop the war. If blame belongs anywhere, it should be placed on a system of compartmentalized government in which nobody, apparently, had simultaneous knowledge of the intelligence *and* on the technical development and capability of the atomic bomb. It is most urgent that governments appoint, for peace and war, a group of persons, below the presidential level, who have military, technical, and intelligence information available to them, and who have enough knowledge to evaluate this information, as well as enough time and freedom from routine work to make a thoughtful analysis and come to a reasoned conclusion.

Jungk describes the atomic scientists' spontaneous movement for civilian control of atomic energy in the U.S., and for international control through a supranational agency. I believe Oppenheimer's role in the latter endeavor is not sufficiently appreciated by Jungk: next to Niels Bohr, who had the original idea, probably Oppenheimer made the greatest contribution to the American plan. One may still regret that this generous plan was not accepted by the U.S.S.R.

There followed the "bitter years" in which all the enormous effort of scientists to steer atomic energy into peaceful channels away from bombs, seemed in vain. Looking back on it, the effort was perhaps not quite futile: The statements we then made now seem to be generally accepted and form part of the public consciousness, both in the West and in the East; statements like "There is no defense against atomic weapons" and "All-out nuclear war will destroy civilization," have become almost platitudinous. It is true, no definite steps have been taken to make such a war impossible, but at least we have survived 13 years without one, more than many of us expected.

Some of Jungk's remarks about the "bitter years" (1947–1955, I suppose) are greatly exaggerated or even false. According to Jungk, all of us lived in constant fear of being falsely accused of subversion, of having our clearance withdrawn, of being deprived of the means to work. True, all this happened to some scientists, and under ugly circumstances. However, the picture of scientists being constantly under the Damocles sword of public disgrace is certainly false; and though we were concerned about the fate of the few,* most of us worked quite undisturbed on our research.

It is also misleading to stress as much as Jungk does, the warnings (around 1946) by some scientists against the increasing support of science in the U.S. by military agencies. I believe even the then prophets of doom would agree that science has flourished under this support, that it has not become "enslaved" by the military, that support continued to go to pure science while military projects were kept entirely separate, and that especially the Office of Naval Research has an enviable record.

Jungk is undoubtedly correct in saying that the uncompromising, hostile attitude of the Russians, beginning with their rejection of the Acheson-Lilienthal plan, through such events as the Berlin blockade, and on to Korea, was the most important reason for the return of many atomic scientists to the weapons laboratories. The Russian success in developing the A-bomb as early as 1949, earlier than even atomic scientists had predicted, gave added stimulus: We were truly in an armaments *race*, as had been predicted in the Franck Report of 1945 and often thereafter. One should not forget that probably our emphasis on atomic weapons in turn stimulated and accelerated the work in Russia.

Jungk's description of the H-bomb development is remarkably correct; here, as in other parts of his book, he must have done very earnest research and talked to many of the people concerned. As far as I can judge, both the

---

*There were many protests by scientists, individually and in groups, against the abuses of "security." In contrast to this, the German university professors, with a few notable exceptions, did not protest when their colleagues were dismissed in 1933, even though at that early time this would not have involved any great danger to them, especially if it had been done by large groups.

motives and the actions of the various actors in the drama are correctly presented. There were, of course, many important persons, and especially many technical contributors, who are not mentioned.

Jungk considers the Russian bomb exploded in August 1953, "more advanced" than the American "Mike" of November 1952. This assessment is almost certainly false. But Jungk can hardly be blamed for this opinion since it was held by many U.S. officials—in part, presumably, to justify their original contention, in late 1949, that Russia had probably started H-bomb development already, before our own decision to go ahead.

In June 1951, Teller had just developed his new idea on the design of an H-bomb which led to success during the next year. The idea was presented by Teller at a small meeting in Princeton, and was acclaimed by all the physicists present. This meeting is described quite well by Jungk, on the basis of evidence presented during the Oppenheimer hearing of 1954, but he severely criticizes Oppenheimer for changing his opinion about the H-bomb because the project now looked, in Oppenheimer's words, "technically sweet." This, Jungk says, shows "a dangerous motive" of the modern scientist, that technically sweet projects are irresistible, even if he has to sell his soul to the devil.

For some scientists this may be correct. But the main argument on this occasion was that the H-bomb had now become inevitable. My desire, and that of many others, to prove that it could not be made, had not been fulfilled. With its feasibility proved, and the strongest government support behind it, the bomb would undoubtedly be accomplished and most important, it had now become likely that the Russians, sooner or later, would be able to make it too—and all of us felt, with General Omar Bradley, that it would be intolerable to have this weapon in the hands of the Russians but not in our own.

Jungk gives a fair, quite moderate account of the Oppenheimer trial. My opinion about the trial is well known and has been best stated by Wernher von Braun in testimony before a Congressional committee: "In England, Oppenheimer would have been knighted."

According to Jungk, scientists should feel responsible for the political consequences of their work, that they should consider these before they start work on anything as destructive as an atomic bomb, and that they should, if necessary, refuse, as a body, to do such work. Although the first two points have considerable appeal, I cannot agree with the last point. It would set the scientific community up as a superpolitical body. Clearly, the political judgment of scientists as a body, as well as that of individual scientists, is fallible, just as much as professional politicians. In addition, scientists will disagree on political matters, just like other groups. Setting the scientist up as the sole judge of his actions means, in the extreme, condoning treason such as that of Klaus Fuchs—whom, indeed, Jungk does not seem to condemn.

Yet I also disagree with the opposite point of view, quite popular in the McCarthy era, summarized in the slogan, "Scientists should be on tap, not on top." I don't believe we have completely overcome this attitude, and I am afraid it is occasionally shared by scientists, including some who seem to be rather "on top" themselves. According to this view, scientists should only speak when asked, and should confine themselves to reporting scientific and technical facts; they should not express their own opinion, nor make moral judgments of the kind the General Advisory Committee made in their famous H-bomb session of October 1949.

I believe this attitude to be wrong; Jungk's opposite attitude—that the scientist should act exclusively on his own conviction—is a healthy antidote even though in itself also wrong. I believe it is not only the privilege but the duty of the scientist, individually and collectively, to make his opinion and vision known to the government—but it is for the government, not the scientist, to make the last decision, which may well turn out contrary to his expressed opinion, and he should then abide by it. Reporting of facts alone by the scientists, excluding any opinion, might be right if decisions were really made by "the people." But the people unfortunately have far too little knowledge of the issues to make the decision. Actually, decisions are made frequently by a few very busy people in the government to whom the problem is entirely new. The scientist, on the other hand, has usually lived with the problem on which he is being consulted for a long time; he can often visualize future developments as, for instance, when the Franck Report predicted the armaments race.

There is a further argument why the scientist should make his nontechnical opinion known to the government in matters of weapons. The agencies of government which are best informed in these matters are necessarily military departments, professionally obliged always to be in favor of the strongest, most destructive weapons technically attainable, and must be against any measures of disarmament. The only other group of people who are fully informed are scientists working on weapons. Since they don't have, a priori, a professional interest one way or the other, they should be able to consider nonmilitary factors, political as well as ethical ones. Their opinion is therefore valuable in arriving at a balanced decision.

In order to fulfill the function of contributing to the decision-making process, scientists (at least some of them) must be willing to work for government and in government, and they must be willing to work on weapons. They must do this also because our present struggle is (fortunately) not carried on in actual warfare, which has become an absurdity, but in technical development for a potential war which nobody expects to come. Scientists must preserve the precarious balance of armament which would make it disastrous for either side to start a war. Only then can we argue for, and embark on, more constructive ventures like disarmament and international cooperation which may eventually lead to a more definite peace.

Thus my view of the "social responsibility" of the scientist is rather the opposite of Jungk: It is no longer possible for all scientists to return to the "golden years" of the 1920s, to the ivory tower of pure research. Some of us can do so, and should do so, to the extent that "big science" *can* live in the confines of an ivory tower. Others have to work on weapons, much as they know their terror and abhor their actual use, and have to make their knowledge and opinion known to their government, much as politics may give them a feeling of frustration and divert them from science.

# Ultimate Catastrophe?

In the November 1975 issue of the *Bulletin of the Atomic Scientists*, H. C. Dudley claimed that it is possible that an atomic weapon might ignite a thermonuclear reaction in the atmosphere (or the ocean) and thus destroy the Earth. This claim is nonsense.

Dudley's claim is based on a report by the writer Pearl Buck of an interview with Arthur H. Compton in 1959. Had Dudley consulted the original literature, or any of the scientists conversant with the problem, he would soon have recognized that Pearl Buck had completely misunderstood Arthur H. Compton.

The basic theory of thermonuclear reactions in the atmosphere was developed in a report by E. J. Konopinski, C. Marvin, and E. Teller, published by the Los Alamos Laboratory. This work was done before the first nuclear test at Alamogordo in July 1945, and assured the leaders and members of the project that no danger existed that the test would set the atmosphere on fire. Report LA-602 was circulated in August 1946; it was secret until February 1973, when it was declassified.

The only "candidate" for a possible thermonuclear reaction in air was a collision between two nuclei of $N^{14}$ because $N^{14}$ is a nucleus of relatively high energy content. Several nuclear reactions are possible, of which we mention

$$N^{14} + N^{14} = Mg^{24} + He^4 + 17.7 \text{ MeV} \tag{1a}$$
$$= O^{16} + C^{12} + 10.6 \text{ MeV}. \tag{1b}$$

Reaction (1a) has the advantage that the energy of the products is well above the Coulomb potential barrier so that the product nuclei can emerge from the reaction without any difficulty. Reaction (1b) has the advantage that it is not necessary for the two nitrogen nuclei to come in contact at all, but it is only necessary for a deuteron, $H^2$, to "leak" from one nitrogen nucleus to the other. This makes (1b) the more likely reaction, especially because the deuteron is much lighter than the $N^{14}$ nucleus. Konopinski considers reaction (1a), and assumes that *all* nitrogen nuclei which penetrate the Coulomb barrier will undergo this reaction. We shall, later on, also consider reaction (1b) which was not well understood in 1945.

Konopinski assumed that, in some manner, the air could be heated to a (nuclear) temperature of 1 MeV (million electron volt), which is equivalent to 11 billion degrees (Kelvin). Even at this fabulous temperature, the energy produced by the nuclear reaction would be only one one-thousandth of the energy lost by various physical processes. In order for the nuclear reaction to be sustained, it must produce at least as much energy as is lost by other processes, otherwise the temperature will drop and the reaction comes to an end.

What are these loss processes? In a nuclear reaction, the resulting energy is primarily used to heat atomic nuclei. These need not be in thermal equilibrium with the electrons; therefore, we speak of a nuclear temperature. But the nuclei give some of their energy to electrons by collisions, and thus heat the electrons. The electrons, in turn, lose their energy to radiation (bremsstrahlung). This equilibrium equation was solved by Konopinski, with the result that at 1 MeV nuclear temperature, the electron temperature is about 35 keV (thousand electron volt). The figure of one-thousandth in the last paragraph was calculated on this basis. There are other processes, like the inverse Compton effect, which further increase energy losses, and which are discussed in the Konopinski paper.

So there was never *any* possibility of causing a thermonuclear chain reaction in the atmosphere. There was never "a probability of slightly less than three parts in a million," as Dudley claimed. Ignition is not a matter of probabilities; it is simply impossible.

At that, we have assumed that the air is heated to a nuclear temperature of 10 billion degrees. In 1945, it was already realized that this was very difficult. In the meantime, we have learned a great deal more about thermonuclear reactions. They are much more difficult to ignite, and the temperature existing in the reacting medium is much lower than was assumed in the 1940s. For this reason, the "classical super," working on the reaction D + D, was given up in favor of the idea of Teller and S. M. Ulam.

One of the chief mechanisms working against ignition is the powerful emission of radiation by electrons when they are heated to high temperatures. To overcome such radiation, it is necessary to have a nuclear reaction of very large cross section at low nuclear energy, at least as large as the D + D reaction. The temperatures in fission and fusion weapons are usually of the order of a hundred million degrees, occasionally somewhat higher, but never as much as one billion degrees. The air cannot be heated as hot as the weapon. At one billion degrees, reaction (1b) should be considerably more likely than (1a). We have estimated the energy produced by reaction (1b), making deliberate overestimates. Even so, (1b) fails to supply the energy lost to the electrons by a factor of $10^{16}$!

The safety factor against ignition of the atmosphere is enormous. This will remain true unless, some time in the future, nuclear weapons of entirely different type are designed which produce much higher temperatures —a very unlikely possibility.

What is true for air is even more true for water in the ocean. The nuclear reaction

$$O^{16} + H^1 = F^{17} + \gamma \tag{2}$$

is exceedingly slow, as we know both from the interior of stars and from laboratory experiments. The only reaction which has *a priori* a good probability is

$$D^2 + D^2 = H^3 + H^1 \tag{3}$$
$$\text{or He}^3 + n.$$

But only one hydrogen atom in 6,500 is deuterium (D), so the number of collisions between them is extremely small. And all the light hydrogen and the oxygen would have to be heated as well! The reaction

$$D^2 + H^1 = He^3 + \gamma \tag{4}$$

is improbable *and* suffers from the low abundance of deuterium.

Dudley is aware that for the maintenance of thermonuclear reactions, "the energy released must be retained near the site of its production" and that high pressure is required for this to happen. But he then claims that the pressure required is "tons per square inch," and that such pressures exist at the bottom of the oceans.

In reality, however, the pressure near the center of the sun is of the order of a trillion tons per square inch, and it is such pressures that are required to retain the energy near the site of its production. The pressure in the oceans is totally insufficient for this purpose. Compared to the total pressure in the center of the sun, it is smaller than the pressure of a fly on a tabletop as compared to the pressure at the bottom of the oceans.

In fact, a little retention of radiation energy makes matters still more unfavorable for a nuclear reaction, because of the "inverse Compton effect" in which radiation picks up energy from electrons. This is discussed in detail in the paper by Konopinski.

In reality, one must, of course, discuss how radiation can escape from the region where the energy is produced. This problem is more subtle but the result is the same: the radiation will not be retained sufficiently in the ocean. In the air, of course, it is even worse.

Dudley has nightmares which have no relation to reality. Ignition of the atmosphere or the ocean by atomic weapons is not a matter of probability, but simply will not happen.

There are many excellent reasons against nuclear war, and these are well known to our statesmen as well as to our scientists. On this one point, I can fully agree with Dudley: there must not be nuclear war. But it is totally unnecessary to add to the many good reasons against nuclear war one which simply is not true.

# ARMS CONTROL

# The Case For Ending Nuclear Tests

The cessation of nuclear weapons tests has been debated in the newspapers, in Congress, and most of all, in Geneva, where there have been what I consider honest negotiations. Concrete proposals have been made by both sides, by the East and by the West; there has been a considerable measure of agreement between East and West; and a number of articles of a treaty have been accepted.

The fundamental points of view of the two sides on disarmament measures were quite different. The Soviet Union advocated just a piece of paper on which we would agree to reduce our armaments. The United States and Britain insisted on the verification of any treaty concerning the limitation of armaments.

Because of the Russian desire for secrecy, the West proposed that the verification of a test cessation agreement should be primarily by physical methods, which would mean less intrusion into the privacy of the Soviet Union. It is clear that the Russians have by now accepted the major principle on which the United States has insisted; namely, that there should be a control system for the test cessation agreement. This in itself is an important result of the negotiations, and we must not jeopardize this achievement by either breaking off the negotiations or by making unreasonable demands which we know Russia cannot fulfill.

The problem of detection of nuclear explosions varies with the medium in which the nuclear explosion is set off. Until recently, practically all the nuclear explosions were set off in the atmosphere, either on the ground or higher up. The best-known method for detection of nuclear explosions is the collection of radioactive debris, by planes flying in those regions where the debris is expected to arrive, or on the ground from fallout. However, the collection of fallout on the ground is quite unreliable, because winds may carry the radioactive debris in one direction or another.

We can also detect a nuclear explosion by the acoustic method, which consists in recording the pressure wave created by the explosion. Though the pressure decreases as the wave passes through the atmosphere, it remains recordable to a distance of 10,000 miles. The wave is such a good indicator of nuclear explosions that the United States has usually announced Russian explosions very soon after they have taken place. We have announced far more Russian explosions than the Russians themselves. Similarly, the Russians have been able to detect our nuclear explosions.

To improve detection, it was proposed, as part of a treaty on the cessation of nuclear tests, to have a large number of stations in each country, especially in Russia, in the United States, and in the British Commonwealth, all equipped with detection instruments. With the network of stations worked out in the Geneva negotiations, it was generally agreed that we would be able to detect and identify explosions in air down to the level of one kiloton, or possibly lower.

The second medium in which nuclear explosions have been set off is underwater. Such explosions are, if anything, even easier to detect than those in air, because the pressure wave is very well propagated through water, so much so that even an explosion of a few tons, not kilotons, can be recorded through water for thousands of miles.

Another location where nuclear tests might be carried out is in outer space. Detection there is considerably more difficult than in the air or in water, but it is confidently expected that one could detect nuclear weapons tests to distances of at least a million miles, which is four times the distance from here to the moon.

Underground testing is our most vexing problem and has received the most publicity. It has the obvious advantage that it does not contaminate the atmosphere, and therefore the great disadvantage, from a detection point of view, that radioactive air samples cannot be collected. From experience gained in Nevada, we know how deep an explosion has to be buried in order to prevent escape of radioactive material into the air. A kiloton bomb must be buried about 400 feet underground; a 20-kiloton bomb, about 1,100.

The displacement of the earth produced by an underground explosion is sufficiently great to be recorded easily by a seismograph, unless the explosion itself is very small. The two largest underground explosions to date were carried out in Nevada in the fall of 1958. One of these was 5 kilotons and the other 20; they could be observed on the seismographs throughout the United States, and the larger one gave a clear signal in Russia.

Unfortunately, underground explosions produce the same type of record as earthquakes; namely, seismograms consisting of a series of perhaps 20 wiggles. There are very few ways to distinguish between the two types of seismogram. The best distinguishing mark that seismologists have been able to find is the direction of the so-called first motion—whether the first wiggle starts up or down.

The criteria for distinguishing earthquakes from explosions were discussed in detail in the Geneva negotiations. It was decided that control stations should be set up at regular intervals of about 600 miles in seismic regions—that is, regions in which earthquakes normally occur—and about 1,000 miles in aseismic regions. This comes to about 20 stations in the U.S.S.R.

It is estimated that 100 to 200 earthquakes with a force equivalent to that of an explosion of 20 kilotons occur in the Soviet Union every year. Of these, about half would be distinguished from explosions by first motion and other features in the seismograms obtained in the Geneva network of stations. This leaves about 50 to 100 earthquakes a year which cannot be distinguished from explosions.

The only sure way to tell an earthquake from an explosion is to send an inspection team to the location of the earth disturbance. A combination of seismograms from several stations can determine the location to an accuracy of about five miles. Thus, one would have to send an inspection team to explore an area of about 100 square miles for evidence of an explosion. A number of scientists have tried to work out procedures for such an inspection.

How many inspection teams would have to be sent out every year? Would one inspect every one of the 50 to 100 questionable events? I do not think so. It is generally agreed that about 30 percent would suffice. This would mean about 20 inspections in the Soviet Union annually in order to monitor possible nuclear tests above 20 kilotons.

A detailed study of the problem has been made by Richard Latter of the Rand Corporation. He finds that the capability of the control system would be greatly increased by distributing the stations somewhat differently from the pattern proposed in the Geneva negotiations and by increasing their number in Russia from 20 to 30. This would make it much easier to distinguish earthquakes from explosions, so that only about 10 earthquakes per year in the Soviet Union, with a force equivalent to 20 kilotons or more, will remain unidentified by their seismograms. Ten inspections would therefore cover all doubtful events equivalent to 20 kilotons or more. Forty inspections would make it possible to monitor all earthquakes above 5 kilotons. (A 5-kiloton bomb is a small bomb. The Nagasaki bomb had a force of 20 kilotons.)

We must keep in mind, however, that our underground explosions in the past have been carried on in Nevada tuff, in a rock which is very soft and which therefore gives a relatively strong seismic signal. If the underground explosion is carried out in harder rock, such as salt, granite, or limestone, it creates a smaller signal. A signal which corresponds to a 5-kiloton explosion in tuff may correspond to perhaps 10 or 15 kilotons in these harder types of rock. Even 15 kilotons is not a very large atomic explosion.

Our capability of detection and inspection of underground explosions under the Geneva or the Latter system would be quite satisfactory, were it not for the possibility of deliberate concealment of explosions by a process known as decoupling, or muffling.

A very powerful method has been proposed by Albert Latter, the brother of Richard Latter. His method consists of making an enormous underground cavity and setting off the atomic bomb in the middle of the cavity. One can calculate that the apparent size of the explosion is thereby reduced by a factor of about 300.

Latter's decoupling theory was invented about January 1959, and was then checked by many scientists, including me. It was experimentally verified with small explosions of conventional high explosive in Louisiana early in 1960. To decouple the explosion of a 20-kiloton weapon, a spherical hole of about 50 million cubic feet, or nearly 500 feet in diameter is necessary. Moreover, the hole has to be 3,000 feet below ground. The big room of Carlsbad Caverns is only big enough to muffle 10 kilotons.

It would obviously be very slow and very expensive to excavate such a hole by a normal mining operation, with pick and shovel and high explosives. However, in salt domes, large holes can be made by washing out the salt; that is, by pumping water in and pumping brine out. Experts of the oil industry have estimated that to excavate a hole big enough to muffle a 20-kiloton explosion would take more than two years and would cost about $10 million. This is quite a lot of money, but the time factor is probably more important. The actual washing operation has to be preceded by an engineering study, and after completion of the hole, considerable time is required to set up and complete an atomic test.

Whether a 500-foot hole in a salt dome could be maintained is not known. Holes of about one-tenth that volume are used to store oil and gasoline. It is also unknown whether a hole could be used repeatedly for nuclear testing.

I cannot imagine that the washing of a 500-foot hole would go undetected, even in a closed country like the Soviet Union. Just the amount of salt water which would be carried up to the earth's surface is staggering, being many times the volume of the salt. Somehow, the salt water must be disposed of. One possibility is to dump the brine directly into the sea or into depleted oil wells, but both of these methods severely limit the geographical areas in which the hole could be constructed. A more widely applicable method would be to dump it into a river. But even in a very big river, like the Ohio or the St. Lawrence, the salt excavated from one big hole would double the salt content for a year. The increase would be very easy to discover by making a chemical analysis of the river water at regular intervals.

It is important that there are relatively few locations in which large holes can reasonably be dug; one needs salt domes and, in addition, ways to dispose of the salt water. In other types of rock, excavation and disposal of the excavated material would be far more troublesome, and the resulting hole would probably be less safe and less reliable for decoupling. Luckily for inspection, the geographical regions where salt domes are known to exist are generally aseismic, so that any seismic signal originating from these regions would be suspicious. For the same reason, only a few inspections would be needed to guard against muffled shots. Of course, one first has to get a signal from a muffled explosion. How this can be done I shall discuss later.

The main question is: Does any country want to go to such an extreme as constructing the big hole in order to cheat on a test ban? Can we really assume that the Russians would go to the trouble of negotiating a test cessation treaty just in order to turn around the next day and violate it?

Having participated in the negotiations with Russian scientists at Geneva on three occasions, I believe that they are sincere in wanting the test cessation agreement and do not intend to cheat. For instance, in November 1959, although the Russians were in many ways reluctant to agree with the American delegation, they were very eager to accept any improvements in detection apparatus suggested by the Americans. If the Russians wished to violate the treaty, they would have objected to these improvements.

Many other Americans, without disagreeing on any of the scientific facts, believe that the Russians are bent on violation, and therefore they oppose cessation of testing. Edward Teller has argued this point repeatedly, and in particular in his television discussions with Lord Bertrand Russell. His line of reasoning runs as follows: "We cannot detect Russian underground tests of bombs of small yield; since we cannot detect these tests, we should assume that they are carrying out such tests. If they carry out such tests and we do not, then they will soon be ahead of us in the area of small nuclear weapons. When they are ahead of us in this area, they will have military superiority, and they can blackmail us into complete submission. At that moment the free world will have to capitulate to Russian Communism."

It seems to me this is a series of *non sequiturs*. Every one of these steps, I think, is very unlikely. I do not think the Russians intend to violate a treaty banning weapons tests; I do not think that the Russians could risk cheating, even if there is only a small likelihood of being detected. Even if we had no system of physical stations detecting nuclear tests, the Russians would not risk having some defector tell us about a clandestine nuclear explosion. If there were such a defector telling us of a Russian violation, it would not be very difficult to find physical evidence of it. I believe that the Soviet Union, which is posing as a peace-loving nation, whether rightly or wrongly, simply cannot afford to be caught in a violation, and therefore I think that it will not try to cheat.

But even supposing that the Russians wanted to cheat, what would they need to do? One violation, one nuclear test below ground, does not do much for the development of weapons. In the first place, they would have to develop entirely new methods to assess the results of a nuclear test. Most of the methods commonly used for observing the results of a test in air do not work underground. Probably several explosions of weapons whose performance is already known are required to develop methods of observation underground. Second, two or more test explosions are often needed to develop a single new weapon. Finally, a country which already has dozens of types of weapons will hardly be interested in developing just one more, in violation of a treaty. For all these reasons, a potential violator of the treaty would only be interested if he could perform a whole series of tests.

Now, if a series of tests were carried out, all at the same location, this would greatly simplify the work of the detecting agency. It would merely be necessary to detect the disturbances on the seismograph, not to distinguish them from earthquakes. Repeated seismic disturbances, originating from the same location (except as aftershocks of one big earthquake), would be sufficiently suspicious to warrant dispatching an inspection team to their site. With our accepted Geneva system of 20 stations in the Soviet Union, it is generally agreed that we could record disturbances underground of less than one kiloton. Thus, to violate the treaty without being detected, it would be necessary to find a new location for practically every test in the series. The development of a new test site would add enormously to the cost, complexity, and inconvenience of testing.

I had the doubtful honor of presenting the theory of the big hole to the Russians in Geneva in November 1959. I felt deeply embarrassed in so doing, because it implied that we considered the Russians capable of cheating on a massive scale. I think that they would have been quite justified if they had considered this an insult and had walked out of the negotiations in disgust.

The Russians seemed stunned by the theory of the big hole. In private, they took the Americans to task for having spent the last year inventing methods to cheat on a nuclear test cessation agreement. Officially, they spent considerable effort in trying to disprove the theory of the big hole. This is not the reaction of a country that is bent on cheating.

Two of the Russian scientists presented to the Geneva Conference their supposed proof that the big hole would not work. A day or two later, Latter and I gave the counterproof and showed, with the help of the Russian theory itself, that the Russian proof was wrong, and that the theory of the big hole and the achievable decoupling factor were correct. We have been commended in the American press for this feat in theoretical physics. I am not proud of it.

In hearings before the Joint Atomic Energy Committee of Congress, Teller recommended that we in the United States should continue determined research to find out further methods of decoupling, further methods of reducing the signal from an underground explosion. His argument is that we have to know all the possible methods of concealment if we are to develop a detection system which can deal with them. This may be so, but should we really spend our time and effort drawing up a blueprint for a violator of the treaty, and also do the engineering development for him?

The Russians themselves have been quite consistent in their attitude toward decoupling. In negotiations in Geneva for joint research on improvement of seismic detection, they refused, in May 1960, to participate in any research or decoupling or to permit the United States to engage in such research. "The Russian people," said Mr. Tsarapkin, the U.S.S.R. delegate, "will not understand it if research under the test ban treaty is conducted for the purpose of defeating the treaty."

Because of the difficulties of detecting small nuclear tests underground, President Eisenhower on February 11, 1960 proposed to the two other great nuclear powers to effect, for the time being, only a partial test ban treaty. Tests in the atmosphere and underwater, as well as in the nearer parts of outer space, would be discontinued. He proposed to ban all large nuclear explosions underground, those giving a signal equivalent to 20 kilotons or more under Nevada conditions. Smaller explosions and decoupled tests would be permitted, because they cannot be identified by the present system.

Further, the president proposed, the three powers should start intensive research on the improvement of methods for detection and identification of underground explosions, so that in time the treaty could be extended to smaller underground explosions, and perhaps a complete ban could be effected in the end.

This proposal was accepted by the U.S.S.R. on March 19, 1960 with an important modification; namely, that there should be a moratorium on smaller nuclear tests for a number of years. While only the large tests would be prohibited in the treaty itself, the three powers would declare in a separate document that they would refrain from carrying out nuclear explosions under 20 kilotons. The Russians proposed four to five years as the duration of such a moratorium; the West, two to three years.

To pursue this proposal further, the seismologists of the three nuclear powers met in May 1960 in Geneva to make plans for joint research on seismic detection. A large measure of agreement was achieved, but there is still a question as to the number and type of nuclear explosions to be used in this research.

Somehow, the seismic research has become identified in the mind of the public with the setting off of underground test explosions. This is by no means correct. The main problem is to improve the instruments of detection. This can be done largely by utilizing signals from earthquakes. Few explosions will be needed.

We need instruments which will give us more diversified information. We want to eliminate from the seismograms as far as possible the "noise"; that is, the ever-present, minute, irregular motions of the earth. We need to learn to utilize calculating machines and mathematical tricks for the analysis of the records. We want to learn to deduce the shape and depth of the original earth displacement caused by a disturbance hundreds or thousands of miles distant from the recording instrument. At present, the depth can be deduced from the seismogram only if it is very great—about thirty miles or more. Perhaps we can learn to determine the depth of the source of the seismic signal to an accuracy of one mile. If this were possible, then signals originating from a depth of at least two miles could be attributed to earthquakes, because it is extremely difficult to drill holes that deep.

A number of promising ideas for improved detection were proposed in the spring of 1959 by the Berkner Panel, which was set up by the President's Science Advisory Committee to study the problem of detection of nuclear explosions. Some of these ideas were proposed to the Russians at the Technical Conference in November 1959. With a lot of research there is good hope for substantial improvement of the art of detection during a two- or three-year moratorium.

Would it be possible by these improvements to detect a decoupled test of a 20-kiloton explosion in a big hole, using only the 20 seismic stations provided in the Geneva system? We do not know. It must be remembered that, with a decoupling factor of 300, a 20-kiloton explosion looks like only 70 tons, not kilotons, and this is a very small explosion indeed. With the present methods, the Geneva system can only detect and locate explosions down to about 700 tons, not identify them.

But suppose it is impossible to improve the seismic techniques sufficiently by research. We already know one method which would certainly enable us to detect even the fully decoupled tests. This method is to make the spacing between stations much smaller. My proposal would be to decrease the spacing from 600 miles, as agreed to in the Geneva negotiations, to 120 miles in the seismic regions, and also in those regions where there are salt domes. In the parts of the Soviet Union where there are neither earthquakes nor salt domes, the stations could be distributed at a much wider spacing—let us say, 250 miles. With such spacing, one would need about two hundred stations to cover the Soviet Union.

If the whole of the Soviet Union, seismic and aseismic regions alike, were to be covered by stations at 120-mile spacing, 600 stations would be needed. I mentioned this number during the congressional hearings which were held in April 1960, under the chairmanship of Mr. Holifield. The number was then quoted out of context and without the proper qualifications in sensational press reports which implied that I no longer supported a test ban treaty. This, of course, did not correspond to the intention of my testimony.

The 20 stations in the U.S.S.R. which are provided by the Geneva system will be large, manned stations, each with about 30 technical people plus supporting personnel. They will have equipment to observe nuclear tests in the air as well as underground, and each station will have an array of 100 seismographs in order to reduce the "noise."

The additional 200 stations which I am proposing would be unmanned robot stations with one seismograph each, or possibly two. Such a system of robot stations would be simpler, cheaper, and, at the same time, more effective for seismic detection than the 20 large stations. I should think that these additional robot stations might well be acceptable to the Russians, especially if we do not demand them now but keep this idea in mind as a way out if no better method is discovered to observe decoupled tests.

How much would such a system of small stations cost? Two estimates have arrived independently at a figure of about $100,000 for each of these small stations. This includes provision for making the station "tamper-proof." While one cannot expect such tamperproofing to be 100 percent effective, one can expect to design the station in such a way that any tampering will be observable.

It is important to have reliable transmission of information from the robot stations to the large stations. Various methods of transmission have been considered by a group of communications experts. the total cost of the system of small stations in Russia, even with the most elaborate communication links, is estimated to be less than $200 million. Engineering estimates have also been made of the cost of the basic 20-station net. No final figure has been given, but a total of $500 million for both the small and the large stations is probably the best conservative guess that can be made at present. This is not cheap, but it is a great deal less than $1–5 billion for the large stations alone which was mentioned in the Holifield hearings in April 1960.

The system of robot stations is expected to detect and identify fully decoupled 20-kiloton tests in the seismic regions. It is expected to detect and locate, but not identify, similar tests in the aseismic regions. I have previously shown that, once you get a signal, the number of inspections can be held within reasonable bounds. Of course, the robot stations would also greatly improve the detection of normal, not decoupled shots. It would be easy to detect and identify earthquakes down to 1 kiloton or less. Hence, the number of inspections could be greatly reduced.

It has been argued that several years will pass before we can build the system of small robot stations. This is true, but it would also take several years before a sufficiently large cavity could be constructed and successfully used.

In the meantime, there is a possibility of using already existing, smaller cavities in salt. There seem to exist, at least in this country, several holes which would fully decouple an explosion of 2 or 3 kilotons. If similar cavities exist in Russia, it must be admitted that it would be possible to use these for testing very small weapons; there will always be a threshold of detection, even with the best detection system. The question is whether such testing would be worthwhile. Moreover, a 2-kiloton hole could be used to achieve partial decoupling for a larger explosion—say, 20 kilotons. From the Louisiana experiments, one may estimate that this proportion of hole size to explosion energy will give decoupling by a factor of 30. The 20-kiloton explosion then looks like two-thirds of a kiloton. This is just detectable by the Geneva net of stations, although without the robot stations a signal of this size could not be distinguished from an earthquake. Thus, with the Geneva net alone, it would be risky for a violator to attempt a partially decoupled explosion of 20 kilotons. A 10-kiloton explosion, partially decoupled in an existing 2-kiloton hole, could probably go undetected until either we have robot stations or significant improvements have been made in the art of detection. However, it seems from the Louisiana experiments that the cavity will suffer damage when it is used for partial decoupling of an oversized explosion. It is, then, very doubtful whether it could be used more than once. In this way, the few existing big holes would soon be used up.

I believe, therefore, that it is technically feasible to devise a system of detection stations and inspections which will give reasonable assurance against clandestine testing, with the possible exception of very small, decoupled tests.

A key point in the monitoring system is on-site inspection. This was recognized as necessary in the Experts Conference of the summer of 1958. The Russians agreed reluctantly, but they did agree, to a passage which says that "all events which are recorded by the control stations and which could be suspected of being nuclear explosions will have to be inspected on the site." This passage, which our negotiators insisted on, has been a very powerful argument for our side. I am sure that the Russians have often regretted that they agreed to it. On the other hand, our delegation would not have agreed to the final report of the conference without this passage.

Estimates made in Washington have ranged as high as from 100 to 300 necessary inspections per year in the Soviet Union for a complete test ban. Obviously, the Soviet Union wishes to prevent such a large number of inspections. Soviet negotiators have objected time and time again that we would use the on-site inspections for espionage purposes.

To solve this impasse, Prime Minister Macmillan, on his visit to Moscow in the spring of 1959, proposed that there should be a fixed quota of inspections every year in each country and that, in exchange, each side should have free choice of the inspections to be made on the territory of the other.

In other words, if a suspicious event is recorded on the seismographs in Russia, then the West, which means essentially the United States and Britain, would have the right to choose whether or not to inspect this event. They would decide on the basis of other knowledge whether it is reasonable in that particular locality to suspect a nuclear explosion. If, for example, the seismic event is shown to originate from a completely trackless wilderness in the mountains, then it may be presumed by the West that it was not a nuclear explosion. If, however, the seismic event is recorded in an abandoned mining area where one might very easily dig a tunnel to put in a nuclear bomb for testing, or especially in a salt dome area, then we would presumably insist on inspecting that area.

It seems to me that this idea of a quota, combined with free choice by the West of the events to be inspected in Russia, is a good compromise. Russia has accepted this idea in principle; our government has not: it insists on inspection of a percentage of the suspicious events rather than a fixed number. This implies that the number of inspections would be determined on the basis of technical need. The Russian position is that the quota must be determined on the basis of political acceptability alone.

In practice, however, the U.S. government has adopted the quota idea as part of President Eisenhower's proposal of February 11, 1960, of a partial ban on nuclear tests. He proposed that this partial test ban be enforced by a quota of about 20 inspections per year in the U.S.S.R., a very reasonable number. The Russians, while accepting the February proposal with some modification, have not as yet responded to the proposed number of inspections. In fact, they have never officially mentioned any number and have given us to understand only that they are thinking of very small numbers of inspections. But one point is clear: The quota must be fixed by the negotiators before either Russia or the West will sign a treaty.

L et us assume the most unlikely and worst possible case; let us assume that the Russians have gone to all the trouble of negotiating a treaty only in order to violate it. It would take then a very long time to set off any significant number of explosions; it would take a tremendous effort. It is generally agreed that they could test only small nuclear weapons. Even in the area of small nuclear weapons, a test series would take a number of years. Now, let us even assume that the Russians wish to go to all this trouble just to develop further small nuclear weapons. Where do we stand?

At this time, it is generally agreed that we are far ahead of Russia in the development of small nuclear weapons. We have nuclear weapons ranging from 20 kilotons down to a fraction of a kiloton. We have them in all different sizes. We have weapons which can be carried in big airplanes, in fighter planes, in ballistic missiles, in land-based rockets, and even in airborne rockets to bring down enemy planes. We have nuclear weapons which can be shot in short-range rockets, like Honest John; we have nuclear weapons so small that they can be carried by the infantry with relative ease. We have an enormous arsenal of such weapons. The Russians also have a number of such weapons, but their arsenal in the field of small weapons, as far as we know, is much more limited than ours, and probably their weapons are not quite as sophisticated. It would take them a long time to catch up with us.

What is the alternative? Suppose we resume tests only in the area of small weapons. Then we could be sure that it would not take the Russians very long to reach our present, very high-level technology in this field. But, it will be argued by Teller and his associates, in the meantime we also can make progress. Clearly, we could if we resumed nuclear tests. However, we have already gone far enough so that very little we can do in the future will be of great military significance.

While the Russians could gain considerably by the resumption of tests of small nuclear weapons, they also have enough such weapons to give them a sizable capability in case of a tactical nuclear war. Therefore, they do not have a desperate need for improving their weapons, and thus not enough incentive for testing to risk a violation. Yet, if nuclear tests were resumed legally, the Russians would probably make more rapid progress than we would.

It has been claimed that we still have a long way to go in nuclear weapons. There are two schools of thought on this: one essentially represented by the Livermore Laboratory, the other by the Los Alamos Laboratory. The Los Alamos Laboratory, our chief laboratory in the development of nuclear weapons, which has to its credit most of the weapons which have been developed up till now, is generally of the opinion that not much more can be obtained in the way of weapons improvement. They will admit, and I will admit, that there can always be some further improvement, but the question is: Is it worthwhile? Even if testing were allowed, would it be worth a great effort?

Let me repeat: If the Russians really want tactical nuclear weapons— that is, nuclear weapons of small yield—then the best thing for them to do would be to resume testing of such small weapons officially, exactly as was suggested in the original proposal by President Eisenhower on February 11, 1960. The fact that they asked instead for a moratorium on small tests indicates to me that they do not put much weight on development of these weapons.

I believe that if tests were resumed, if the negotiations on test cessation were to break down, then the Russians would choose to test big weapons, hydrogen weapons, in the megaton class. I think that these are the weapons which are most dangerous to us; these are the weapons which the Soviets consider to be of the greatest military importance to them. I want to remind you of the situation in the fall of 1958. At that time the United States put in an extra test series, which is known as the Hardtack II series, carried out in Nevada. It consisted of a number of small nuclear explosions, from about 20 kilotons down to a fraction of a kiloton.

At the same time, the Russians tested the largest weapons they have ever tested, weapons of many megatons. They increased the yield from their weapons; they made them more efficient in the large-megaton class. They probably obtained in that test series the warhead for their big inter-continental ballistic missile. This is a tremendous weapon in their hands. If the Russians were to launch a determined attack on this country, perhaps a surprise attack, they would undoubtedly use these enormous weapons.

I believe that it was much to our disadvantage that the United States con-ducted the test series in the fall of 1958. It would have been in our power to declare in August of 1958 that we would stop testing from that moment on. If President Eisenhower had made this bold declaration, I do not believe that the Russians would have conducted their test series in September and October. They used our small test series as a very welcome excuse to con-duct their own. The bold political decision to stop tests altogether would have been much more to our military advantage, since such a decision would have made it difficult for the Russians to resume testing and to test their big warheads. Had the Russians tested nevertheless, we would still have been free to conduct the small Hardtack II series. We got compara-tively little from our small test series; they got a great deal, and very big weapons, out of theirs.

If we were now to resume testing, if we were to break off negotiations in Geneva, then we could not force the Russians to test only small weapons underground. While we might restrict our own testing in this manner, the Russians could test anything that they considered to be to their military advantage. Whether their tests would be in the multimegaton category, or whether they would try to decrease the weight for, let us say, a one-megaton weapon, I could not predict. But I strongly suspect that they would choose to test large weapons rather than small ones, that they would choose to test strategic weapons, which they consider the most important, rather than the tactical weapons which have had such prominence in public discussions in the United States.

If we had stopped nuclear testing when the Russians first suggested doing so, at the beginning of 1956, we would presumably have had a very great superiority in hydrogen bombs. We had tested at least half a dozen; they had tested one type only. We might possibly have a situation in which the Russians would not now have a hydrogen warhead for their ICBM. The missile gap has been a great worry to our military in the late fifties. Without a hydrogen warhead, the ICBM would be much less important, and our superior position in planes might have remained of much greater significance.

The politically bold decision to accept the Russian offer to stop nuclear testing in 1956, either before or after our own 1956 test series, would then have given us considerable military advantage. Sometimes insistence on 100 percent security actually impairs our security, while the bold decision—though at the time it seems to involve some risk—will give us more security in the long run.

There can be no doubt that, since 1956, the Russians have gained in nuclear weapons, relative to us. It is my belief that this is quite natural: the country that is behind will catch up; the country that is ahead will not make so much progress in the future. Teller has pointed out that our nuclear weapons today are about a thousand times as efficient as they were in 1945. He states, "In comparison with the nuclear weapons of 1960, those of 1950 appear completely obsolete. If the development should continue, there is no doubt that in 1970 nuclear explosives can be produced compared to which our present weapons will appear similarly outdated." The first sentence is clearly true. But if we want to increase the efficiency of our nuclear weapons by another factor of about ten—not a thousand—from the presently achievable, we come to a point where the entire material in the weapon must undergo a nuclear reaction. Since there must be assembly mechanism, triggers, bomb cases, and the like, this is clearly impossible. Further nuclear weapons development will be limited by the laws of physics.

This being so, further testing by both sides would bring the Russian capability closer and closer to ours. If we stop nuclear testing now, we may reserve at least the little bit of military advantage in nuclear weapons that we still possess. It is certainly late enough. So I come to the conclusion that, even from the purely military point of view, for our purely military strength compared with Russia's, we would gain by a test cessation agreement.

The political gain would be enormous. Basic to the accepted control system of a test ban are the control stations on the territory of the contracting parties. If the agreement becomes reality, the Russians will for the first time permit foreign international inspectors to go on their soil, to have the right to check up on their activities. This is the first time that the Russians have been willing to give up any part of their sovereignty. Of course, we are asked to give up the same part of our sovereignty too, but for the Russians, with their extreme desire for secrecy, it is a far more difficult thing to do, and it would be a real achievement of the negotiations.

The Russians have further admitted that there should be on-site inspections of suspicious events detected by the physical control stations. This is another major concession. So, in the test cessation agreement, we would get the first admission in principle of the rights of a foreign control organ on Russian soil, an admission which might be of the utmost importance for further disarmament agreements. It would be very dangerous indeed for us to jeopardize this achievement by not concluding the test cessation agreement.

The main importance of our negotiations on the test cessation agreement comes, I believe, not from this agreement itself, important as it is, but from further agreements which must follow. It has been recognized widely in the United States, and also in the Soviet Union, that the continued arms race makes no sense. The two countries are fully capable of destroying each other, in fact, of destroying each other several times over. This is an absurdity. Modern war simply does not make sense as an instrument of national policy. I believe that we should try to arrive at a situation of carefully controlled limitation of armaments. It is a difficult thing to achieve, and it will be a long road before we do.

However, if we want to stop the armaments race, then we have to make a start somewhere. It has to be made in a way consistent with United States policy, meaning that every limitation of armaments must be carefully controlled. We have to make a start in an area where it does not cost us too much, where we can back off again if the first treaty doesn't work. With the test cessation agreement, this would be possible, because it covers a sufficiently restricted subject.

I have so far discussed this problem entirely in regard to the two powers, the United States and Russia. However, the effect on other powers is at least equally important. It is clear that other powers may also get nuclear weapons soon. In fact, if we wait long enough, they surely will do so. I do not know how long it will take for China to achieve this capability. I do not think it will be next year, but I will not be surprised if the Chinese develop nuclear weapons before another five years pass. It is in our interest to keep nuclear weapons out of China's possession. We have every interest in restricting the nuclear club to its present members, essentially three, with France a junior partner. If the three great nuclear powers continue nuclear testing, then there will be no pressure on the other countries to refrain from developing nuclear weapons. If, however, the three great powers give up nuclear testing, and give it up completely, then popular pressure, both from the great powers and from the small powers, will be very strong on the other countries to make them adhere to the treaty which the great powers have signed.

I cannot predict whether China will in fact adhere to a test cessation treaty. I think that both Russia and the United States would desire that China do so. But one thing seems certain to me: If we do not have a treaty on cessation of nuclear tests, then China will surely get nuclear weapons in a fairly short time. For this reason, it is imperative not only to have a treaty but to have it soon.

At this time we can still get something if we agree to stop nuclear testing. But we have a wasting asset here. Before long, I believe, public opinion in the world will force us to stop nuclear testing without our getting anything in exchange. At present we get in exchange recognition by Russia of stations on Russian soil and of the principle of controlled disarmament. We may further get in exchange the restriction of the nuclear club to three members.

Opponents of the test cessation agreement want to have a perfect agreement; they want to have an agreement in which we can be sure to detect each and every violation, no matter how small. I think that by insisting on perfection we shall end up with nothing.

# Disarmament and Strategy

In the fall of 1961, the United States was deeply stirred by the Russian atomic weapons tests and by the debate on fallout shelters. Nuclear war suddenly seemed very real. I believe one should not shrink from contemplating the realities of nuclear war but I also believe that, through the particular development of strategic weapons during the last few years, nuclear war is becoming less probable every year. This development, tending toward invulnerability, will create much greater military stability and is a good starting point for meaningful disarmament negotiations.

The need for disarmament, for a halt to the arms race, was dramatically illustrated by the Russian test of an atomic weapon of 60-megaton yield in October 1961. Shortly before the test, Khrushchev announced that by testing a weapon of 50 megatons, the Soviet Union would prove at the same time the design of a weapon yielding 100 megatons.

One-hundred megatons is a formidable weapon. The strongest effect of such a mammoth weapon is to set fires, and this effect is increased by exploding the weapon at high altitude, let us say, 30 miles. The book, *Effects of Nuclear Weapons,* published by the Atomic Energy Commission (AEC) and the Department of Defense in 1957 and revised in April 1962, indicated that on a clear day wooden houses and forests may be set on fire by such an explosion to distances of about 30–50 miles from ground zero, and highly inflammable materials, like loose paper, oil-soaked rags, etc., to greater distances. It is quite possible that a firestorm will also result but the conditions for firestorm have never been clearly determined. On the other hand, such an explosion at high altitude will not cause appreciable fallout; this will only be produced by explosions at, or near, the ground.

Blast will be small from an explosion 30 miles high. But if the explosion is set off at only a few miles elevation, dwelling houses will be destroyed by blast up to about 20 miles away. It is clear that one such weapon could wipe out any one great population center.

From the military point of view, there is not too much difference between weapons of 100 megatons or 10. Ten-megaton weapons have existed for a long time, and 10 megatons is enough to destroy any big city. At 10 megatons, the most important destructive mechanism is still probably blast. In the AEC weapons effect handbook, a 10-megaton weapon, exploded close to the ground, will give a pressure of three pounds per square inch at a distance of nine miles, severely damaging wooden frame houses and brick buildings. Reinforced concrete buildings are able to stand several times this pressure. Assuming a destructive radius of 10 miles, a 10-megaton bomb, exploded at the southern end of Central Park, would destroy all of New York City except for a few outlying districts, like Coney Island. Maybe it is good that the Russian test has finally made the world really aware of the mortal peril in which we live.

In 1950, after the first Russian tests of fission bombs, the United States thought that terror must be met by greater terror, A-bombs by H-bombs. As could have been expected, the Russians also developed H-bombs. Thus, a balance of terror was created, and both sides adopted a strategy of deterrence: neither side can damage the other without suffering terrible retaliation. The horrible weapons then exist only in order that the other side does not use them. I do not see how this precarious situation can be changed until these arms are eliminated by enforceable disarmament.

Under the pressure of the Korean War, with its heavy loss of American lives, John Foster Dulles formulated "the doctrine of massive retaliation"—i.e., Communist attacks on the periphery of the free world would from now on be answered by atomic bombs on Russian or Chinese territory. This is a most pernicious doctrine because it means deliberate escalation of a small war into a big one. It was, I believe, also a shortsighted doctrine because it was already clear in 1954 that Russia would soon get the H-bomb.

The massive retaliation doctrine has now been turned back on us. After the U-2 incident, Khrushchev threatened to attack with rockets, presumably with thermonuclear warheads, any country which would serve as a base for U-2 flights over Russia. Thereby, he adopted the same doctrine of escalation of a small act into a big war. Recently, he went further by threatening to use his 100-megaton bomb against the West in case of war.

We thus see that both Eastern and Western statesmen use the word retaliation for two entirely different concepts. One is retaliation against an enemy attack by H-bombs. Our capacity to retaliate in such a case is vital to us, it is the only *military* force which prevents a potential enemy from making an attack. The other is massive retaliation against minor wars. This concept, I believe, must be removed from our military doctrine, both for practical and for moral reasons.

With both sides in possession of H-bombs, it is obviously tempting for a country to try to remove the retaliation capacity of another country—to destroy the enemy's war-making capability at the outset of war. This is known as the strategy of counterforce. In the mid-1950s, when planes were the only vehicles capable of delivering large bombs, destruction of the enemy's strategic planes was a great prize. If a surprise attack succeeds in destroying enemy planes on the ground, clearly the deterrent no longer exists; it has failed. The country striking first wins the war.

There are two possible defenses against such a calamity. The more obvious one is air defense—the attempt to shoot down invading planes by fighter planes or anti-aircraft rockets. These forms of air defenses were developed by both sides; in fact, the Russians used a much larger fraction of their military budget than we did. It is not now clear to what extent their defenses, or for that matter, ours, can be penetrated by invading planes.

The second measure against surprise attack is to put one's strategic bombers on alert. For this and other purposes, such as civil defense, we constructed a radar warning system, known as the DEW (Distant Early Warning) line, which is strung out along the northern edge of the North American continent. It would give two or three hours warning time of approaching enemy planes. In case of warning, our strategic bomber force would take off and fly north, loaded with bombs. Since a radar warning can be caused by all sorts of things besides enemy planes, such as a flock of birds or certain kinds of radio noise, the force is under instructions that it may not proceed beyond a certain line without a specific command from the president of the United States. This measure is known as fail-safe and I hope the name is justified.

Thus, defense against a surprise attack by planes has gradually been strengthened by air defense and by the alert. However, this situation was changed again drastically in favor of the aggressor with the advent of the ICBM, the intercontinental ballistic missile. When the Russians tested their first ICBM in the summer of 1957, they tipped the scale heavily in their favor. They were so far ahead of U.S. development that many Americans, both military and civilian, could not believe the intelligence reports until Sputnik appeared in the sky. Sputnik greatly accelerated our military research and development, especially our own development of ICBMs so as to reestablish the balance of strategic forces. Further, Sputnik induced great improvements in the organization of science by the U.S. government at the presidential level and in the Department of Defense.

But while its indirect effects on the United States may have been beneficial, the main consequence of the ICBM was to make a surprise attack much easier. We developed a radar warning system called BMEWS, ballistic missile early warning system, but its warning is not very early. It takes about 15 to 20 minutes from the time BMEWS sees a missile until the time the missile arrives in the United States. Fifteen minutes is better than nothing and, accordingly, the Strategic Air Command (SAC) has many bombers on ground alert capable of take-off on a 15-minute warning. Some bombers are actually in the air all the time. In this way we are protecting, at least in the interim, our strategic retaliatory capability.

Our own first operational missiles are just about as "soft" as planes—they are above ground so that an H-bomb dropped many miles away will destroy them. This vulnerability tempts the enemy to use counterforce, and encourages our own strategists to get our missiles off before they are destroyed, in other words, to shoot on the basis of radar warning alone. Even though radar engineers constantly attempt to minimize false alarms, to my knowledge they have never been completely eliminated. Clearly, it is far more dangerous to launch a missile attack than to let the planes take off. Planes can be recalled, missiles cannot.

It will be seen that, coupled with the highly mechanized pattern of our systems necessary to achieve desired rapidity of response, the counterforce concept has become extremely dangerous. Any accident may trigger the nuclear holocaust. Even the most peaceful nation may be induced to start a war if information is received indicating that the enemy is about to attack—and this information may be wrong. Once war starts, each side will be inclined to use a maximum of striking power, because whatever is not used immediately may be destroyed by an enemy strike in the next hour. Thus the situation is highly unstable, accidental war is not unlikely, and if war comes it is immediately full scale.

Moreover, if you expect perhaps only 10 or 20 percent of your force to survive the beginning of the war, you must provide for a force which is five or ten times as large as what you think you will actually need. We have a force of about 500 B-52 heavy bombers and about 1,000 B-47 medium bombers. The Russian forces, presumably, are smaller but still formidable. Let's assume a B-47 carries only 10 megatons and a B-52 20, then bombs carried on planes alone total 20,000 megatons. And since 10 megatons will destroy any large city, it is quite clear that these forces are far beyond anything that can reasonably be considered necessary for war.

Counterforce strategy puts a premium on the first strike; it thus makes war more likely, and increases the level of strategic forces beyond reason.

The military problem of the last few years then was to make the retaliatory force survive an initial attack. The solution which is just being put into effect, is to put missiles, especially Minuteman and Titan, into hardened underground silos which can survive a blast pressure of at least 100 pounds per square inch, probably more. An explosion of 10 megatons causes a pressure of 100 pounds per square inch at a distance of one and a half miles. Therefore, if silos are spaced more than three miles apart, it is impossible by one attack of 10 megatons to destroy two Minutemen at the same time. In other words, it would take at least one 10-megaton warhead to eliminate one "tiny" Minuteman, or several smaller warheads, aimed with an accuracy at present probably not obtainable. The precise location of all these silos would have to be known to the attacking enemy, very unlikely, even in the "open" United States. The difficulty is compounded by the fact that at least at present the distance between points in the U.S. and the U.S.S.R. is apparently not known to as good an accuracy as one mile. It is reasonable to assume then that hardening to 100 pounds per square inch and proper spacing between the Minuteman and Titan silos will make our launch sites essentially invulnerable to enemy attack. The situation would not be changed if 100-megaton weapons were used: it might perhaps be possible to destroy two little Minutemen with such a large missile. It seems clear that 100 missiles of one megaton each constitute a far stronger force than one missile of 100 megatons.

Our Minuteman is propelled by solid fuel which makes it instantly ready for take-off. The first tests of Minuteman have been very successful. I believe that in spite of our later start on intercontinental missiles, we now are building a more useful missile force than the Russians have at present. Of course, they may change their force. According to news reports at present, we also have an edge in the number of missiles actually deployed—a grim note of hope that we have a force to prevent a rational enemy from undertaking thermonuclear attack.

Even more invulnerable than the Minuteman is the Polaris missile (also propelled by solid fuel). To be safe from counterattack by Polaris submarines, an enemy would have to find all, or nearly all, of them: it is notoriously difficult to locate submarines. We now have Polaris submarines capable of launching about 100 missiles. With Polaris and the hardened Minuteman, the premium for a first strike is essentially removed; the danger of surprise attack is greatly reduced, and so is the instability of the world. If the Russians should also develop, and come to rely on, a similar type submarine, they might be less nervous and this would benefit our interests too. Polaris and Minuteman together offer us a rather secure second strike force. Of course, any force can be used for a first strike, but a first strike by small missiles is limited. If a country is mainly interested in first strike and aggression, the large 10- and 100-megaton warheads will be more effective than small ones, and it will not need to make its missiles and planes invulnerable. On the other hand, if we reduce our force essentially to a second strike force, it will be one of the best demonstrations that we are only interested in defense and not in first attack.

Until now the Russians have sought to protect their strategic force by secrecy. If we do not know the location of their airfields and missile sites, obviously we can not launch a successful counterforce strike. In their recent nuclear test series, however, there may be some indication that they may adopt methods more similar to our own in the future. The test series included many in the range from about one to five megatons, just the range which might be suitable for a Russian second strike missile in hardened sites. I hope that this is the correct interpretation of this phase of their test series and that the Russians are also aiming for an invulnerable second-strike force, not a first-strike force. In this case, their test series may well have reduced the danger of war.

The fact that the striking force can be made secure against surprise attack should make it possible to greatly reduce its size. As already noted, the large size of our present strategic force is justified mainly by the expectation that only a fraction of the force would survive surprise attack. When missiles are essentially invulnerable, you need much fewer and it becomes possible to slow down the arms race, even without a formal arms-limitation treaty. However, in order to feel safe in limiting the size of its forces, each side must feel sure that the other side does not have vastly superior forces. One 10-megaton weapon, if properly aimed, might destroy one of our Minuteman sites. If the number of large Russian ICBMs were to exceed greatly the number of our hard missile sites, then we would have to increase the number of our missiles as well. It is very important, therefore, that both sides observe restraint by deploying small missiles primarily and limiting their number. In the past, the Soviet Union has followed a policy of deploying only limited numbers of planes and missiles. With a fairly good knowledge of their strength, we can limit the strength of our forces. Left in the dark, we must prepare on the assumption of the worst possible case. This is one strong reason why Soviet excessive secrecy is very damaging to the cause of peace, and is often against their own interests. In the past, when we received information on Soviet military strength from U-2 flights, it often led to limiting our own military preparations and, thus, indirectly reduced world tension.

Another desirable consequence of the invulnerable missile site, and of Polaris, is that secrecy loses most of its value. If the sites can not be destroyed by available numbers of atomic weapons, then it is no longer useful to know where they are. This should make it possible for ourselves, and especially the Russians, to greatly reduce secrecy. I believe that openness should be possible in a world armed with invulnerable missiles and not only in a disarmed world, as Khrushchev has said.

The concept of the invulnerable missile force has sometimes been attacked as a Maginot-line philosophy. There are decisive differences: From a hardened missile launch site, you can counterattack the enemy anywhere—even in his homeland. From a Maginot fortress, you cannot. Once a Maginot line is breached, the remaining fortresses are useless in stemming the enemy advance. (In fact, the Maginot line was not breeched but circumvented because it was incomplete at its western end.) Each Minuteman silo can act on its own, even if others are destroyed. Of course, just as in the case of the Maginot line, technological changes might make missile silos vulnerable.

It will always be necessary to know the changing threat from the potential enemy and to adjust the structure of strategic forces to the changes. For instance, recent Soviet tests may require certain modifications in the details of our weapon-systems military planning, even if they have not changed the fundamental soundness of our plans for an invulnerable missile force.

But the invulnerable deterrent has at least one disadvantage. As long as the strategic force consists of planes, which can easily be destroyed, they will draw enemy fire. In fact, many strategic analyses have assumed that if there is a war, then the fire will be concentrated entirely on military targets. But once these military targets are effectively invulnerable, then to be destructive an enemy will probably concentrate on bombing cities. This conclusion is most unfortunate and against the sound military tradition of trying to destroy the enemy's war-making capability. But I believe we have no choice but to accept this drawback of the stable deterrent.

The invulnerable deterrent clearly has an important effect on civil defense. We have just had a long and bitter debate about fallout shelters. Some widely publicized articles argue that once each of us has a shelter we can await thermonuclear war with calm and 97 percent of us will survive. Others, somewhat less loudly, claimed that fallout shelters were totally useless. A few took a middle course, as did the useful pamphlet on shelters recently published by the Department of Defense.

Whether shelters are useful or not depends, of course, on the kind of attack against which they must protect. If the attack is exclusively on military targets, if the bombs are all exploded on the ground, and if there is only one major strike,then fallout shelters will be very useful. But this implies an extreme counterforce strategy on the part of the enemy which, as I have explained, is highly unlikely because it is ineffective against an invulnerable missile force. If the attack is exclusively by means of 100-megaton weapons exploded at high altitude, then fallout shelters are totally useless because there will not be any fallout near the explosion. (There will only be worldwide fallout, months later, which is at a sufficiently low level so that shelter protection is unnecessary.) The reality, if war should come, will probably lie somewhere in between: there will be some fallout and fallout shelters will save many lives, especially in smaller cities and in the country. But fallout shelters will be useless in those big cities which will be actual targets. Further, their value anywhere is diminished if there are repeated enemy strikes, which is likely if both sides have invulnerable forces. Thermonuclear war remains a terrible catastrophe no matter how extensive is a shelter program.

One objection to the shelter program is that its greatest effectiveness is in the framework of a counterforce strategy from which we are trying to get away, that it tends to make the American people accept thermonuclear war as a reality when we are trying hard to make such a war less and less likely. A shelter program is acceptable only as an insurance against the possibility that we may be unsuccessful, insurance which may or may not pay off. To avoid a nuclear holocaust must remain the main goal of our military policy and this end is served by the invulnerable deterrent force.

Some military experts have stated that the stable deterrent is likely to remain stable until one side or the other finds an effective civil defense or an antimissile missile, the so-called AICBM. I think it is clear that any really effective civil defense is impossible and I believe the same is true of AICBM. This proposition is more difficult to prove, largely because much of the argument is classified. The difficulty is not the most obvious one, that it is difficult to hit a fast-incoming missile by your own antimissile. This can be done; in fact, we have announced some successful tests of our antimissile, the Nike Zeus. There is also no problem about providing suitable atomic warheads for antimissiles, contrary to the claim of many opponents of the nuclear test ban.

The offense has always many more possibilities than the defense. The offense can choose its target, it can concentrate a lot of fire on one target while the defense has to defend 20, 50, or 100 different targets. The offense can always bring in enough missiles to use up all the defensive missiles which may exist near one target. The offense can fire a salvo of many missiles simultaneously which will saturate the radars. And most important, the offense can send, with the actual missiles, decoys, gadgets which look to the radars just like missiles. In spite of intensive work on this problem it remains extremely difficult to find any way to tell them apart. For these reasons, I believe there is no effective AICBM system.

I have discussed how the stable deterrent decreases the likelihood of thermonuclear war drastically because a nation no longer needs to attempt to be the first to strike. This recognition makes it possible for nations to use restraint. It is no longer necessary to respond to confusing radar signals by launching missiles. Even an accidental attack by one enemy missile or bomber need no longer trigger a war, but can be answered by suitable diplomatic action. But the most important restraint can be used in war itself: It is no longer necessary to shoot off clouds of missiles for fear the enemy will otherwise destroy them, but it is possible to limit the number actually fired to that which appears necessary. Thus I believe that, even if the deterrent should fail, a war in the era of invulnerable missiles need not lead to complete destruction. On the other hand, this means, of course, that such a war will not necessarily be over after one exchange of thermonuclear weapons, in contrast to the present belief associated with the counterforce concept. Each side is capable of destroying the other virtually completely, inflicting casualties which experts have put variously at 25, 50, or 90 percent of the population, and even in the case of relatively moderate casualties, destroying the fabric of society and its institutions.

Once a deterrent force is essentially invulnerable, it would even be possible to adopt some scheme such as that proposed by Leo Szilard in *The Voice of the Dolphins:* In case of a serious conflict between two countries, one of them may threaten to destroy just one city of the other to which the second may respond by a similar threat. Although the plan may seem strange at first sight, it would certainly mean a great reduction of the level of war from the present one, and it may not be unrealistic in the era of stable deterrent. It must be realized that war can be limited today only by an act of will on both sides. We must learn not to use our full strength. This is frustrating, as will be well remembered from the Korean War, but it is the only way to avoid utter catastrophe. The destructiveness of modern weapons no longer permits unconditional surrender.

To use restraint after war breaks out, though militarily safe, may be psychologically almost impossible, even if governments on both sides are entirely rational. An irrational government like that of Hitler may even launch an attack in spite of the invulnerable deterrent. In both these circumstances, it would help greatly if the level of armaments were much lower than today. This objective could be done by gradual, voluntary reductions on both sides. But it appears much safer to limit armaments by a treaty, strictly enforced by inspection.

Obviously, the breakdown of the test-ban negotiations is not a good omen for disarmament. After three years of negotiations, no treaty was concluded and the Russians resumed testing. However, we should not accuse the Russians of breaking an agreement by their resumption of tests.

The moratorium was no agreement, but a voluntary undertaking on both sides. In fact, the United States repeatedly insisted that the moratorium was temporary and could be terminated any day.

Nevertheless, I believe that the U.S.S.R. showed bad faith. It is very likely that they had started specific preparations by March 1961, when the test-ban conference reconvened in Geneva. Therefore, they negotiated for at least six months in bad faith. They did so just at the time when we were showing most clearly by our concrete proposals at Geneva that we were sincerely interested in a test ban and willing to meet more than halfway Soviet political demands on the control system. Moreover, the kind of weapons tested show that their laboratories had probably been working full speed during the whole moratorium on the assumption that tests would at some time be resumed. Our laboratories put their main emphasis on such improvements as could be used and go into stockpiles without a test. The very gradual start of our testing is the best proof that we had not anticipated resumption.

Many opponents of the test ban, impressed by the technical progress made by the Russians in their 1961 tests, argue that we should never have entered these negotiations. I would draw the opposite conclusion: It would have been a great advantage to us to conclude a test-ban treaty at the earliest moment because it would have stabilized the technical advantages we had in 1958. If we had been less concerned about the perfection of the inspection system and about the possibility of the Russians' concealing some tests in the kiloton range, it might have been possible to conclude a treaty in 1959 and thus prevent their very impressive multi-megaton tests of 1961. Of course, even if we had had fewer reservations on inspection, a treaty might not have been concluded. But if the Russians continued to make difficulties in the face of a more yielding attitude on our part, we might have known their real intentions earlier.

We do not need to look far afield for reasons why the Russians resumed testing. They knew as well as we did that they were technically behind in 1958. By testing in 1961 they were bound to approach our technology more closely even if—as they had to assume—we would answer by tests of our own. The laws of physics put a limit on the explosive yield which can be obtained from a given weight. Both the U.S.S.R. and the U.S. are rapidly approaching this limit, therefore the country that is behind is bound to gain more than the more advanced country, no matter how hard the latter may try. There is thus no reason at all to assume that the Russians thought they could leapfrog over us by their test series; the expectation to approach us more closely was motive enough. There is even less reason to think that the Russians needed clandestine tests before their open ones in order to be sure that their megaton tests would actually work—it is perfectly straightforward to prepare megaton tests purely by theoretical calculations and to expect them to work; we have often done so in the past. Indeed, as far as I know, there has not been any persuasive evidence, despite claims to the contrary, that the Russians conducted clandestine underground tests during the moratorium.

An interesting sidelight is the appraisal of the effectiveness of underground tests. During the negotiations, opponents of the test ban claimed that it was easy to carry out almost any test underground, and that underground testing would not appreciably slow down weapons development. Even to construct big, "Latter" holes in which explosions could be decoupled, was supposed to be easy and quick, at least for the Russians. Now the same group tells us that underground tests are much too slow, that atmospheric tests are needed to make sufficiently rapid progress in our weapons development.

I do not want to pursue these technical problems because I believe it was a great mistake during the negotiations to concentrate so much on technical devices. Everyone, qualified or not, was discussing whether you could detect underground nuclear explosions and distinguish them from earthquakes. There was hardly any consideration in this country of the political implications and of the military advantages or disadvantages of a test ban. One of the best results of negotiations on the test ban was that we recognized our deficiencies and that we set up the United States Arms Control and Disarmament Agency, an autonomous agency loosely affiliated with the State Department. This agency will deal with all the political, military, and technical problems of disarmament. It is one of the most hopeful events in disarmament.

The test ban as an isolated issue, separate from a more comprehensive disarmament agreement, in my opinion, is no longer a desirable goal to pursue. I had two main intentions when I advocated it: to stabilize the technical advantages in nuclear weapons which the United States had in 1958, and to obtain an inspection system which could be a precedent and an example for inspection of future disarmament agreements. Neither of these aims can any longer be fulfilled. The Russians have largely caught up with our technology and they now refuse to permit any elaborate inspection system for such a limited objective as a test ban. Therefore I believe it was quite reasonable for the West, on reconvening in January 1962, to accept the Russian proposal in the spring of 1961 to merge test-ban negotiations with the more important general-disarmament negotiations. However, the Russians no longer liked their own proposal and the negotiations were therefore broken off after 353 sessions.

On the other hand, in the heat of the battle, the value of nuclear tests has been greatly exaggerated. The art of nuclear weapons design is already well advanced. It is generally considered desirable to make weapons lighter for a given yield in order to facilitate their delivery; but only a rather modest decrease in weight is foreseeable. The laws of physics completely rule out a repetition of the development from the Hiroshima bomb to the present in which the energy yield increased by a factor of 1000 for about the same weight. Modest improvements are possible of course but they are certainly not a matter of life and death for the security of the United States. Improvements in the art and reliability of missiles seem to me of far greater military importance.

In the absence of a test-ban agreement, and after the extensive Russian test series in which they attempted to catch up to our technology, it would seem reasonable that we develop nuclear weapons along with other military technology—neither neglecting nor frantically overemphasizing them. In particular, we should test those designs which fit into our strategic plans and have been developed theoretically in the laboratory, in order to be sure that we can rely on our designs and thus on our invulnerable deterrent. There are some weapons effects tests which we might profitably make. But I do not believe that nuclear testing is the endless frontier that some people seem to see in it.

If the test ban is no longer a desirable field for disarmament, then what is? One area which has been discussed is a cutoff of production of fissile material and a reduction of the nuclear weapon stockpile. A cutoff of production is attractive because it would be rather easy to inspect production plants. Still, it is not very effective, because the United States and the U.S.S.R. have enough fissile material to supply a huge strategic force and more. A reduction of the stockpile of nuclear weapons, which might be an effective disarmament measure, would be more difficult to inspect because nuclear weapons are rather small. They can be stored almost anywhere and are hard to find, even if one has complete access to a country. The best way to monitor such a reduction would probably be to inspect the records of past production in each country to try to trace the material which has been produced.

To my mind, the most promising and the most important area in which to start disarmament is that of strategic-delivery vehicles, as proposed by de Gaulle a few years ago. Atomic weapons don't mean much unless they can be delivered to the enemy country. Because of their size, large bomber planes and long-range missiles are easy to find, once you have some access to a country. In the era of invulnerable deterrent, it is no longer necessary to have large numbers of such strategic-delivery vehicles. We now have about 500 B-52s and 1000 B-47s. In a few years we are also likely to have about 1000 invulnerable long-range missiles. It seems to me entirely safe to reduce this varied force to a few hundred missiles, once they are invulnerable, each carrying perhaps one megaton, in contrast to the 10 or 20 megatons now carried by our planes. This effect would be an enormous reduction of the total destructive force. It may be useful in this connection to remember that all the bombing raids on Germany in World War II together added up to one megaton. But this drastic reduction will only be possible if we can be sure that the other side has made a similar reduction.

This brings us to the all-important problem of disarmament— inspection. In September 1961, in a very important document, John McCloy, the disarmament adviser to President Kennedy, and Mr. Zorin of the Soviet Union agreed on some principles of disarmament. Among them, specifically and explicitly, was that any disarmament treaty should include inspection. Though the two sides did not agree on the kind of inspection, the inclusion of the word in the agreement offers hope, and it is most remarkable and encouraging that the agreement could be concluded in the midst of an intensified phase of the cold war. The Russian offer so far has been to let us witness the destruction of whatever missiles or planes they may in the future agree to destroy. Unfortunately, this is not of great interest to us. We must rather know and inspect the armaments remaining. With their type of inspection we could not guard against new weapons production nor against a false declaration of their initial armaments.

Our type of inspection becomes increasingly important the farther disarmament progresses. Assume, for example, that a country initially has 50 percent more armaments than it declares. This does not make too much difference. But when half of their initially declared armaments have been destroyed then they would actually have twice as much as declared, and when their armaments have been supposedly reduced to one-quarter of the present amount, they would have three times what they are supposed to have. Inspection of production facilities would help with the problem of building new armaments, but again there is the question of completeness of information.

We must admit that the Russians also have good reasons for their point of view. As long as most of their delivery vehicles are soft, as long as they rely on long-range planes, and on above-ground missiles, we could destroy these by surprise attack, and I believe the Russians are as afraid of surprise attack as we are. Indeed, the United States Air Force has espoused the principle of counterforce strategy for so many years that the Russians have every reason to be afraid of letting us know the location of their forces. It is true that new counterforce strategy seems to be getting less popular in the

Defense Department, but how can the Russians feel sure that we have given it up? In this situation naturally they wish to preserve one of their greatest military assets—secrecy of the location of most of their striking force.

Thus we have an impasse: We need to know how many delivery vehicles the Russians have. In other words, we need to inspect, while the Russians need secrecy, at least until they have constructed a completely hardened force of missiles. The best way out of the impasse has been suggested by Professor Louis Sohn of Harvard Law School. He suggests inspection by an ingenious sampling procedure: Each of the two countries, Russia and the United States, makes a map of its own territory, dividing it into a number of areas, let us say 20. Each country chooses the location and shape of these areas to its own liking but in such a manner that they have approximately the same military importance or value. Then the maps are exchanged. At the same time, each country declares the total number of its missiles and airplanes, its important military production facilities, etc., and also the number of each category located in each of the 20 areas. Now Country B, having received all the information from Country A, will choose an area to be inspected, for example, Area No. 17. Then Country B has the right to inspect Area 17 of Country A in all detail, using planes, automobiles, going into factories, and so on, with limitations as stipulated in the treaty. Country A similarly makes an arbitrary choice in Country B and inspects that area.

What does this accomplish? It means that only 5 percent of the country is open to inspection to begin with, leaving the Russians their cherished secrecy for still 95 percent of their country and hence presumably 95 percent of their strategic force, since they presumably will have divided the country into strategically about equal areas. At the same time, we can verify the initial Russian declaration of existing armaments in Area 17 in every detail. It would clearly be extremely dangerous to cheat in the initial declaration, certainly in any major way, because there is always some intelligence information which would tell Country B if some area of Country A has an excessive force of missiles, and Country B would be sure to choose just that area for inspection. Hence, gross inaccuracies in the initial declaration are clearly against the interest of each country, while minor inaccuracies would not disturb the strategic balance. The important point of the Sohn plan is that it checks the accuracy of the initial declaration, and hence establishes that trust between the nations which is so lacking today, and so important for disarmament.

In addition to the method of inspection, a disarmament agreement must, of course, specify the schedule of arms reduction. Let us assume strategic armaments are to be reduced by 10 percent every half year, to be repeated nine times until only 10 percent of initial armaments remain. Every half year, a new declaration of total armaments would have to be made, both by Russia and the United States, which must show at least 10 percent less total force than the previous one. Also every half year another area is picked for inspection by choice of the other country, and again it can be verified that in that particular area the number of delivery vehicles has actually been reduced to the number in the new declaration. The first area presumably remains subject to inspection but this could be arranged differently by agreement.

So far, I have considered the Sohn plan as it applies to the United States and Russia, and to a reduction in the number of strategic delivery vehicles. But of course disarmament must extend to other countries, to the Warsaw Pact, and NATO, to the People's Republic of China, and finally to all countries of the world. Disarmament must also be extended to other weapons categories, naval vessels, fighter planes, tactical nuclear weapons, tanks, guns and standing armies, and so on. It is, of course, important to find out whether such a plan would be acceptable to the countries concerned.

It seems important to me that disarmament, whether along this plan or any other, should not interfere with the establishment of invulnerable forces, neither by us nor by the Russians. Since these forces are only now being built up, this means that presumably the number of weapons in certain categories must be increased while others are being reduced initially. Thus, an agreement should be sufficiently flexible and must not, for example, call for the same percentage reduction in each category. On the other hand, it seems to me essential that negotiations for arms reduction begin now when stable deterrent is in sight. At this point, both sides have reason to feel secure and strong, so that negotiation is from strength on both sides. Postponement to the time when stable deterrent forces are fully developed is unnecessary, if the treaty permits such development. Any postponement is dangerous because new factors, technological changes, or increased power of more nations, may upset the balance. If an understanding between East and West can be reached by that time, such an upset is much less likely to occur.

A drastic reduction of strategic forces with inspection by treaty as suggested by Louis Sohn, seems our best hope for stopping the arms race. To start on this road, both sides will have to change some of their attitudes. The United States will need to renounce its policy of massive retaliation which involves applying a strategic threat against peripheral war. To make it possible, without yielding to the Communists, it will be necessary initially to build up our conventional forces, in line with President Kennedy's idea that we must have a third alternative to nuclear holocaust or surrender. A temporary build-up of our conventional military forces is an essential counterpart to nuclear disarmament. Our young men, as well as those of Western Europe, must be willing to serve in the Army and must understand that this actually decreases the risk of all-out thermonuclear war. For some time to come, we may need to retain tactical nuclear weapons for use in limited wars, even though experts disagree strongly about their usefulness and about whether they favor the offense or the defense. In any case, I believe the important distinction is between strategic war with long-range missiles and tactical war in the vicinity of the battlefield, rather than between nuclear and non-nuclear war. I hope the invulnerable deterrent will be effective enough, so that even if nuclear weapons were used on the battlefield, the war would not escalate into a full-scale strategic nuclear war. When the strategic threat is thus reduced, I expect that ground forces will gain in morale, since this development will greatly raise their importance in our military establishment. Ultimately, I would hope we can eliminate nuclear weapons from tactical as well as strategic forces.

These are some of the things we must do. The Russians, on the other hand, will have to open their country to inspection by a disarmament authority. They will also have to realize that their professed goal of complete and general disarmament can only be reached by means of limited, meaningful steps. Finally, they will need to give up the hope—if they have it—of achieving ultimate military superiority in a continued arms race.

In the McCloy-Zorin Agreement of September 1961, both Russia and the United States agreed to work toward general and complete disarmament, thus accepting the proposal which Khrushchev made to the United Nations in the fall of 1960. Both sides also agreed that this goal can only be achieved in steps, and that no step must put either side at a military disadvantage. There are several other equally reasonable provisions in the agreement.

We must therefore examine what we have to expect of a world in which disarmament is general and complete. First of all, in order to achieve such a disarmed world, we must find ways to settle disputes by other means than war. To guarantee this state of affairs, we must strengthen world law and submit international disputes to final arbitration to the International Court in the Hague. The U.S. does not have a good record in this respect: it is urgent that the Senate repeal the Connelly Amendment which provides that the United States will be bound by decisions of the Hague Court only if it so chooses. Naturally, the Soviet Union also would have to recognize the Court. The recognition of world law would clearly weaken national sovereignty which at present runs counter to the avowed beliefs of the Soviet

Union. To enforce world law, we shall probably need an international police force stronger than the armed forces of any particular country or of any likely alliance. The important question will be: Who controls the international police force? It might indeed be very dangerous if the force were a power of its own, as the Pretorian Guard was in ancient Rome.

These are just a very few of the many problems which must be solved before disarmament can be general and complete. I do not profess to know the solutions. It is clear to me that it will be difficult to reach agreement between the United States and the Soviet Union, especially, and on the organization of a completely disarmed world. Obviously we cannot wait for agreement on all matters before stopping the arms race. It is important to start and to reduce armaments, even if we can not yet eliminate them completely. I believe there are ways to do this which will not endanger, but rather increase, the security of the United States.

# Antiballistic-Missile Systems

## WITH RICHARD L. GARWIN

I
n September 1967, Secretary of Defense McNamara announced that the U.S. would build "a relatively light and reliable Chinese-oriented ABM system." With this statement he apparently ended a long and complex debate on the merits of any kind of antiballistic-missile system in an age of intercontinental ballistic missiles carrying multi-megaton thermonuclear warheads. Secretary McNamara added that the U.S. would "begin actual production of such a system at the end of this year," meaning the end of 1967.

On examining the capabilities of ABM systems of various types, and on considering the stratagems available to a determined enemy who sought to nullify the effectiveness of such a system, we have come to the conclusion that the "light" system described by Secretary McNamara will add little, if anything, to the influences that should restrain China indefinitely from an attack on the U.S. First among these factors is China's certain knowledge that, in McNamara's words, "we have the power not only to destroy completely her entire nuclear offensive forces but to devastate her society as well."

An even more pertinent argument against the proposed ABM system, in our view, is that it will nourish the illusion that an effective defense against ballistic missiles is possible and will lead almost inevitably to demands that the light system, the estimated cost of which exceeds $5 billion, be expanded into a heavy system that could cost upward of $40 billion. The folly of undertaking to build such a system was vigorously stated by Secretary McNamara. "It is important to understand," he said, "that none of the [ABM] systems at the present or foreseeable state-of-the-art would provide an impenetrable shield over the United States. . . . Let me make it very clear that the [cost] in itself is not the problem: the penetrability of the proposed shield is the problem."

In our view the penetrability of the light, Chinese-oriented shield is also a problem. It does not seem credible to us that, even if the Chinese succumbed to the "insane and suicidal" impulse to launch a nuclear attack on the U.S. within the next decade, they would also be foolish enough to have built complex and expensive missiles and nuclear warheads peculiarly vulnerable to the light ABM system now presumably under construction (a system whose characteristics and capabilities have been well publicized). In the area of strategic weapons, a common understanding of the major elements and technical possibilities is essential to an informed and reasoned choice by the people, through their government, of a proper course of action. In this essay, we outline, in general terms, using nonsecret information, the techniques an enemy could employ, at no great cost, to reduce the effectiveness of an ABM system even more elaborate than the one the Chinese will face. First, however, let us describe that system.

Known as the Sentinel system, it will provide for long-range interception by Spartan antimissile missiles and short-range interception by Sprint antimissile missiles. Both types of missile will be armed with thermonuclear warheads for the purpose of destroying or inactivating the attacker's thermonuclear weapons, which will be borne through the atmosphere and to their targets by reentry vehicles (RVs). The Spartan missiles, whose range is a few hundred kilometers, will be fired when an attacker's reentry vehicles are first detected rising above the horizon by perimeter acquisition radar (PAR).

If the attacker is using his available propulsion to deliver maximum payload, his reentry vehicles will follow a normal minimum-energy trajectory, and they will first be sighted by one of the PARs when they are about 4,000 kilometers, or about 10 minutes away. If the attacker chooses to launch his rockets with less than maximum payload, he can put them either in a lofted trajectory or in a depressed one. The lofted trajectory has certain advantages against a terminal defense system. The most extreme example of a depressed trajectory is the path followed by a low-orbit satellite. On such a trajectory, a reentry vehicle could remain below an altitude of 160 kilometers and would not be visible to the horizon-search radar until it was some 1,400 kilometers, or about three minutes, away. This is FOBS: the fractional-orbit bombardment system, which allows intercontinental ballistic missiles to deliver perhaps 50 to 75 percent of their normal payload.

In the Sentinel system, Spartans will be launched when PAR has sighted an incoming missile; they will be capable of intercepting the missile at a distance of several hundred kilometers. To provide a light shield for the entire U.S., about half a dozen PAR units will be deployed along the northern border of the country to detect missiles approaching from the general direction of the North Pole. Each PAR will be linked to several "farms" of long-range Spartan missiles, which can be hundreds of kilometers away. Next to each Spartan farm will be a farm of Sprint missiles, together with missile-site radar (MSR), whose function is to help guide both the Spartans and the shorter-range Sprints to their targets. The task of the Sprints is to provide terminal protection for the important Spartans and MSRs. The PARs will also be protected by Sprints and thus will require MSRs nearby.

Whereas the Spartans are expected to intercept an enemy missile well above the upper atmosphere, the Sprints are designed to be effective within the atmosphere, at altitudes below 35 kilometers. The explosion of an ABM missile's thermonuclear warhead will produce a huge flux of X rays, neutrons and other particles, and within the atmosphere a powerful blast wave as well. We shall describe later how X rays, particles and blast can incapacitate a reentry vehicle.

B efore we consider in detail the capabilities and limitations of ABM systems, let us briefly summarize the present strategic position of the U.S. The primary fact is that the U.S. and the U.S.S.R. can annihilate each other as viable civilizations within a day and perhaps within an hour. Each can, at will, inflict on the other more than 120 million immediate deaths, to which must be added deaths that will be caused by fire, fallout, disease and starvation. In addition, more than 75 percent of the productive capacity of each country would be destroyed, regardless of who strikes first. At present, therefore, each of the two countries has an assured destruction capability with respect to the other. It is usually assumed that a nation faced with the assured destruction of 30 percent of its population and productive capacity will be deterred from destroying another nation, no matter how serious the grievance. Assured destruction is therefore not a very flexible political or military tool. It serves only to preserve a nation from complete destruction. More conventional military forces are needed to fill the more conventional military role.

Assured destruction was not possible until the advent of thermonuclear weapons in the middle 1950s. At first, when one had to depend on aircraft to deliver such weapons, destruction was not really assured because a strategic air force is subject to surprise attack, to problems of command and control and to attrition by the air defenses of the other side. All of this was changed by the development of the intercontinental ballistic missile and also, although to a lesser extent, by modifications of our B-52 force that would enable it to penetrate enemy defenses at low altitude. There is no doubt today that the U.S.S.R. and the U.S. have achieved mutual assured destruction.

The U.S. has 1,000 Minuteman missiles in hardened "silos" and 54 much larger Titan II missiles. In addition, we have 656 Polaris missiles in 41 submarines and nearly 700 long-range bombers. The Minutemen alone could survive a surprise attack and achieve assured destruction of the attacker. In his annual report, the secretary of defense estimated that as of October 1967, the U.S.S.R. had some 720 intercontinental ballistic missiles, about 30 submarine-launched ballistic missiles (excluding many that are airborne rather than ballistic) and about 155 long-range bombers. This force provides assured destruction of the U.S.

Secretary McNamara has also stated that U.S. forces can deliver more than 2,000 thermonuclear weapons with an average yield of one megaton, and that fewer than 400 such weapons would be needed for assured destruction of a third of the U.S.S.R.'s population and three-fourths of its industry. The U.S.S.R. would need somewhat fewer weapons to achieve the same results against the U.S.

It is worth remembering that intercontinental missiles and nuclear weapons are not the only means of mass destruction. They are, however, among the most reliable, as they were even when they were first made in the 1940s and 1950s. One might build a strategic force somewhat differently today, but the U.S. and the U.S.S.R. have no incentive for doing so. In fact, the chief virtue of assured destruction may be that it removes the need to race—there is no reward for getting ahead. One really should not worry too much about new means for delivering nuclear weapons (such as bombs in orbit or fractional-orbit systems) or about advances in chemical or biological warfare. A single thermonuclear assured-destruction force can deter such novel kinds of attack as well.

Now, as Secretary McNamara stated in his September speech, our defense experts reckoned conservatively six to ten years ago, when our present strategic-force levels were planned. The result is that we have right now many more missiles than we need for assured destruction of the U.S.S.R. If war comes, therefore, the U.S. will use the excess force in a "damage-limiting" role, which means firing the excess at those elements of the Russian strategic force that would do the most damage to the U.S. Inasmuch as the U.S.S.R. has achieved the level of assured destruction, this action will not preserve the U.S., but it should reduce the damage, perhaps sparing a small city here or there or reducing somewhat the forces the U.S.S.R. can use against our allies. To the extent that this damage-limiting use of our forces reduces the damage done to the U.S.S.R., it may slightly reduce the deterrent effect resulting from assured destruction. It must be clear that only surplus forces will be used in this way. It should be said, however, that the exact level of casualties and industrial damage required to destroy a nation as a viable society has been the subject of surprisingly little research or even argument.

One can conceive of three threats to the present rather comforting situation of mutual assured destruction. The first would be an effective counterforce system: a system that would enable the U.S. (or the U.S.S.R.) to incapacitate the other side's strategic forces before they could be used. The second would be an effective ballistic-missile defense combined with an effective antiaircraft system. The third would be a transition from a bipolar world, in which the U.S. and the U.S.S.R. alone possess overwhelming power, to a multipolar world including, for instance, China. Such threats are, of course, more worrisome in combination than individually.

American and Russian defense planners are constantly evaluating less-than-perfect intelligence to see if any or all of these threats are developing. For purposes of discussion, let us ask what responses a White side might make to various moves made by a Black side. Assume that Black has threatened to negate White's capability of assured destruction by doing one of the following things: (1) it has procured more intercontinental missiles, (2) it has installed some missile defense or (3) it has built up a large operational force of missiles each of which can attack several targets, using "multiple independently targetable reentry vehicles" (MIRVs).

White's goal is to maintain assured destruction. He is now worried that Black may be able to reduce to a dangerous level the number of White warheads that will reach their target. White's simplest response to all three threats—but not necessarily the most effective or the cheapest—is to provide himself with more launch vehicles. In addition, in order to meet the first and third threats, White will try to make his launchers more difficult to destroy by one or more of the following means: by making them mobile (for example by placing them in submarines or on railroad cars), by further hardening their permanent sites, or by defending them with an ABM system.

Another possibility that is less often discussed would be for White to arrange to fire the bulk of his warheads on "evaluation of threat." In other words, White could fire his land-based ballistic missiles when some fraction of them had already been destroyed by enemy warheads, or when an overwhelming attack is about to destroy them. To implement such a capability responsibly requires excellent communications, and the decision to fire would have to be made within minutes, leading to the execution of a prearranged firing plan. As a complete alternative to hardening and mobility, this fire-now-or-never capability would lead to tension and even, in the event of an accident, to catastrophe. Still, as a supplemental capability to ease fears of effective counterforce action, it may have some merit.

White's response to the second threat—an increase in Black's ABM defenses—might be limited to deploying more launchers, with the simple goal of saturating and exhausting Black's defenses. But White would also want to consider the cost and effectiveness of the following: penetration aids, concentrating on undefended or lightly defended targets, maneuvering reentry vehicles or multiple reentry vehicles. The last refers to several reentry vehicles carried by the same missile; the defense would have to destroy all of them to avoid damage. Finally, White could reopen the question of whether he should seek assured destruction solely by means of missiles. For example, he might reexamine the effectiveness of low-altitude bombers or he might turn his attention to chemical or biological weapons. It does not much matter how assured destruction is achieved. The important thing, as Secretary McNamara has emphasized, is that the other side find it credible. ("The point is that a potential aggressor must himself believe that our assured destruction capability is in fact actual, and that our will to use it in retaliation to an attack is in fact unwavering.")

It is clear that White has many options, and that he will choose those that are most reliable or those that are cheapest for a given level of assured destruction. Although relative costs do depend on the level of destruction required, the important technical conclusion is that for conventional levels of assured destruction it is considerably cheaper for White to provide more offensive capability than it is for Black to defend his people and industry against a concerted strike.

As an aside, it might be mentioned that scientists, newly engaged in the evaluation of military systems, often have trouble grasping that large systems of the type created by or for the military are divided quite rigidly into several chronological stages, namely, in reverse order: operation, deployment, development and research. An operational system is not threatened by a system that is still in development; the threat is not real until the new system is in fact deployed, shaken down and fully operative. This is particularly true for an ABM system, which is obliged to operate against large numbers of relatively independent intercontinental ballistic missiles. It is equally true, however, for counterforce reentry vehicles, which can be ignored unless they are built by the hundreds or thousands. The same goes for MIRVs, a development of the multiple reentry vehicle in which each reentry vehicle is independently directed to a separate target. One must distinguish clearly between the *possibility* of development and the development itself, and similarly between development and actual operation. One must refrain from attributing to a specific defense system, such as Sentinel, those capabilities that *might* be obtained by further development of a different system.

It follows that the Sentinel light ABM system, to be built now and to be operational in the early 1970s against a possible Chinese intercontinental ballistic missile threat, will have to reckon with a missile force unlike either the Russian or the American force, both of which were, after all, built when there was no ballistic-missile defense. The Chinese will probably even build their first operational intercontinental ballistic missiles so that they will have a chance to penetrate. Moreover, we believe it is well within China's capabilities to do a good job at this without intensive testing or tremendous sacrifice in payload.

Temporarily leaving aside penetration aids, there are two pure strategies for attack against a ballistic-missile defense. The first is an all-warhead attack in which one uses large booster rockets to transport many small (that is, fractional-megaton) warheads. These warheads are separated at some instant between the time the missile leaves the atmosphere and the time of reentry. The warheads from one missile can all be directed against the same large target (such as a city); these multiple reentry vehicles (MRVs) are purely a penetration aid. Alternatively, each of the reentry vehicles can be given an independent boost to a different target, thus making them into MIRVs. MIRV is not a penetration aid but is rather a counterforce weapon: if each of the reentry vehicles has very high accuracy, then it is conceivable that each of them may destroy an enemy missile silo. The Titan II liquid-fuel rocket, designed more than 10 years ago, could carry 20 or more thermonuclear weapons. If these were employed simply as MRVs, the 54 Titans could provide more than 1,000 reentry vehicles for the defense to deal with.

Since the Spartan interceptors will each cost $1 million to $2 million, including their thermonuclear warheads, it is reasonable to believe thermonuclear warheads can be delivered for less than it will cost the defender to intercept them. The attacker can make a further relative saving by concentrating his strike so that most of the interceptors, all bought and paid for, have nothing to shoot at. This is a high-reliability penetration strategy open to any country that can afford to spend a reasonable fraction of the amount its opponent can spend for defense.

The second pure strategy for attack against an ABM defense is to precede the actual attack with an all-decoy attack or to mix real warheads with decoys. This can be achieved rather cheaply by firing large rockets from unhardened sites to send light, unguided decoys more or less in the direction of plausible city targets. If the ABM defense is an area defense like the

Sentinel system, it must fire against these threatening objects at very long range before they reenter the atmosphere, where because of their lightness they would behave differently from real warheads. Several hundred to several thousand such decoys launched by a few large vehicles could readily exhaust a Sentinel-like system. The attack with real warheads would then follow.

The key point is that since the putative Chinese intercontinental-ballistic-missile force is still in the early research and development stage, it can and will be designed to deal with the Sentinel system, whose interceptors and sensors are nearing production and are rather well publicized. It is much easier to design a missile force to counter a defense that is already being deployed than to design one for any of the possible defense systems that might or might not be deployed sometime in the future.

L et us now turn to (1) the physical mechanisms by which an ABM missile can destroy or damage an incoming warhead and (2) some of the penetration aids available to an attacker who is determined to have his warheads reach their targets.

Much study has been given to the possibility of using conventional explosives, rather than a thermonuclear explosive, in the warhead of a defensive missile. The answer is that the "kill" radius of a conventional explosive is much too small to be practical in a likely tactical engagement. We shall consider here only the more important effects of the defensive thermonuclear weapon: the emission of neutrons, the emission of X rays and, when the weapon is exploded in the atmosphere, blast.

Neutrons have the ability to penetrate matter of any kind. Those released by defensive weapons could penetrate the heat shield and outer jacket of an offensive warhead and enter the fissile material itself, causing the atoms to fission and generating large amounts of heat. If sufficient heat is generated, the fissile material will melt and lose its carefully designed shape. Thereafter it can no longer be detonated.

The kill radius for neutrons depends on the design of the offensive weapon and the yield, or energy release, of the defensive weapon. The miss distance, or distance of closest approach between the defensive and the offensive missiles, can be made small enough to achieve a kill by the neutron mechanism. This is particularly true if the defensive missile and radar have high performance and the interception is made no more than a few tens of kilometers from the ABM launch site. The neutron-kill mechanism is therefore practical for the short-range defense of a city or other important target. It is highly desirable that the yield of the defensive warhead be kept low to minimize the effects of blast and heat on the city being defended.

The attacker can, of course, attempt to shield the fissile material in the offensive warhead from neutron damage, but the mass of shielding needed is substantial. Witness the massive shield required to keep neutrons from escaping from nuclear reactors. The size of the reentry vehicle will enable the defense to make a rough estimate of the amount of shielding that can be carried and thus to estimate the intensity of neutrons required to melt the warhead's fissile material.

Let us consider next the effect of X rays. These rays carry off most of the energy emitted by nuclear weapons, especially those in the megaton range. If sufficient X-ray energy falls on a reentry vehicle, it will cause the surface layer of the vehicle's heat shield to evaporate. This in itself may not be too damaging, but the vapor leaves the surface at high velocity in a very brief time and the recoil sets up a powerful shock wave in the heat shield. The shock may destroy the heat-shield material or the underlying structure.

X rays are particularly effective above the upper atmosphere, where they can travel to their target without being absorbed by air molecules. The defense can therefore use megaton weapons without endangering the population below; it is protected by the intervening atmosphere. The kill radius can then be many kilometers. This reduces the accuracy required of the defensive missile and allows successful interception at ranges of hundreds of kilometers from the ABM launch site. Thus X rays make possible an area defense and provide the key to the Sentinel system.

On the other hand, the reentry vehicle can be hardened against X-ray damage to a considerable extent. And in general the defender will not know if the vehicle has been damaged until it reenters the atmosphere. If it has been severely damaged, it may break up or burn up. If this does not happen, however, the defender is helpless unless he has also constructed an effective terminal, or short-range, defense system.

The third kill mechanism—blast—can operate only in the atmosphere and requires little comment. Ordinarily when an offensive warhead reenters the atmosphere it is decelerated by a force that, at maximum, is on the order of 100 $g$. (One $g$ is the acceleration due to the earth's gravity.) The increased atmospheric density reached within a shock wave from a nuclear explosion in air can produce a deceleration several times greater. But just as one can shield against neutrons and X rays one can shield against blast by designing the reentry vehicle to have great structural strength. Moreover, the defense, not knowing the detailed design of the reentry vehicle, has little way of knowing if it has destroyed a given vehicle by blast until the warhead either goes off or fails to do so.

The main difficulty for the defense is the fact that in all probability the offensive reentry vehicle will not arrive as a single object that can be tracked and fired on but will be accompanied by many other objects deliberately placed there by the offense. These objects come under the heading of penetration aids. We shall discuss only a few of the many types of such aids. They include fragments of the booster rocket, decoys, fine metal wires called chaff, electronic countermeasures and blackout mechanisms of several kinds.

The last stage of the booster that has propelled the offensive missile may disintegrate into fragments or it can be fragmented deliberately. Some of the pieces will have a radar cross section comparable to or larger than the cross section of the reentry vehicle itself. The defensive radar therefore has the task of discriminating between a mass of debris and the warhead. Although various means of discrimination are effective to some extent, radar and data processing must be specifically set up for this purpose. In any case the radar must deal with tens of objects for each genuine target, and this imposes considerable complexity on the system.

There is, of course, an easy way to discriminate among such objects: let the whole swarm reenter the atmosphere. The lighter booster fragments will soon be slowed down, whereas the heavier reentry vehicle will continue to fall with essentially undiminished speed. If a swarm of objects is allowed to reenter, however, one must abandon the concept of area defense and construct a terminal defense system. If a nation insists on retaining a pure area defense, it must be prepared to shoot at every threatening object. Not only is this extremely costly but also it can quickly exhaust the supply of antimissile missiles.

Instead of relying on the accidental targets provided by booster fragments, the offense will almost certainly want to employ decoys that closely imitate the radar reflectivity of the reentry vehicle. One cheap and simple decoy is a balloon with the same shape as the reentry vehicle. It can be made of thin plastic covered with metal in the form of foil, strips or wire mesh. A considerable number of such balloons can be carried uninflated by a single offensive missile and released when the missile has risen above the atmosphere.

The chief difficulty with balloons is putting them on a "credible" trajectory, that is, a trajectory aimed at a city or some other plausible target. Nonetheless, if the defending force employs an area defense and really seeks to protect the entire country, it must try to intercept every suspicious object, including balloon decoys. The defense may, however, decide not to shoot at incoming objects that seem to be directed against nonvital targets; thus it may choose to limit possible damage to the country rather than to avoid all damage. The offense could then take the option of directing live warheads against points on the outskirts of cities, where a nuclear explosion would still produce radioactivity and possibly severe fallout over densely populated regions. Worse, the possibility that reentry vehicles can be built to maneuver makes it dangerous to ignore objects even 100 kilometers off target.

Balloon decoys, even more than booster fragments, will be rapidly slowed by the atmosphere and will tend to burn up when they reenter it. Here again a terminal ABM system has a far better chance than an area defense system to discriminate between decoys and warheads. One possibility for an area system is "active" discrimination. If a defensive nuclear missile is exploded somewhere in the cloud of balloon decoys traveling with a reentry vehicle, the balloons will either be destroyed by radiation from the explosion or will be blown far off course. The reentry vehicle presumably will survive. If the remaining set of objects is examined by radar, the reentry vehicle may stand out clearly. It can then be killed by a second interceptor shot. Such a shoot-look-shoot tactic may be effective, but it obviously places severe demands on the ABM missiles and the radar tracking system. Moreover, it can be countered by the use of small, dense decoys within the balloon swarms.

It may be possible to develop decoys that are as resistant to X rays as the reentry vehicle and also are simple and compact. Their radar reflectivity could be made to simulate that of a reentry vehicle over a wide range of frequencies. The decoys could also be made to reenter the atmosphere—at least down to a fairly low altitude—in a way that closely mimicked an actual reentry vehicle. The design of such decoys, however, would require considerable experimentation and development.

Another way to confuse the defensive radar is to scatter the fine metal wires of chaff. If such wires are cut to about half the wavelength of the defensive radar, each wire will act as a reflecting dipole with a radar cross section approximately equal to the wavelength squared divided by $2\pi$. The actual length of the wires is not critical; a wire of a given length is also effective against radar of shorter wavelength. Assuming that the radar wavelength is one meter and that one-mil copper wire is cut to half-meter lengths, one can easily calculate that 100 million chaff wires will weigh only 200 kilograms (440 pounds).

The chaff wires could be dispersed over a large volume of space; the chaff could be so dense and provide such large radar reflection that the reentry vehicle could not be seen against the background noise. The defense would then not know where in the large reflecting cloud the reentry vehicle is concealed. The defense would be induced to spend several interceptors to cover the entire cloud, with no certainty, even so, that the hidden reentry vehicle will be killed. How much of the chaff would survive the defense nuclear explosion is another difficult question. The main problem for the attacker is to develop a way to disperse chaff more or less uniformly.

An active alternative to the use of chaff is to equip some decoys with electronic devices that generate radio noise at frequencies selected to jam the defensive radar. There are many variations on such electronic counter-measures, among them the use of jammers on the reentry vehicles themselves.

The last of the penetration aids that will be mentioned here is the radar blackout caused by the large number of free electrons released by a nuclear explosion. These electrons, except for a few, are removed from atoms or molecules of air, which thereby become ions. There are two main causes for the formation of ions: the fireball of the explosion, which produces ions because of its high temperature, and the radioactive debris of the explosion, which releases beta rays (high-energy electrons) that ionize the air they traverse. The second mechanism is important only at high altitude.

The electrons in an ionized cloud of gas have the property of bending and absorbing electromagnetic waves, particularly those of low frequency. Attenuation can reach such high values that the defensive radar is prevented from seeing any object behind the ionized cloud (unlike chaff, which confuses the radar only at the chaff range and not beyond).

Blackout is a severe problem for an area defense designed to intercept missiles above the upper atmosphere. The problem is aggravated because area-defense radar is likely to employ low-frequency (long) waves, which are the most suitable for detecting enemy missiles at long range. In some recent popular articles, long-wave radar has been hailed as the cure for the problems of the ABM missile. It is not. Even though it increases the capability of the radar in some ways, it makes the system more vulnerable to blackout.

Blackout can be caused in two ways: by the defensive nuclear explosions themselves and by deliberate explosions set off at high altitude by the attacker. Although the former are unavoidable, the defense has the choice of setting them off at altitudes and in locations that will cause the minimum blackout of its radar. The offense can sacrifice a few early missiles to cause blackout at strategic locations. In what follows we shall assume for purposes of discussion that the radar wavelength is one meter. Translation to other wavelengths is not difficult.

In order totally to reflect the one-meter waves from our hypothetical radar, it is necessary for the attacker to create an ionized cloud containing $10^9$ electrons per cubic centimeter. Much smaller electron densities, however, will suffice for considerable attenuation. For the benefit of technically minded readers, the equation for attenuation in decibels per kilometer is

$$\alpha = \frac{4.34}{3 \times 10^5} \frac{\omega_p^2}{\omega^2 + \gamma_e^2} \gamma_e.$$

Here $\omega_p$ is the plasma frequency for the given electron density, $\omega$ is the radar frequency in radians per second and $\gamma_e$ is the frequency of collisions of an electron with atoms of air. At normal temperatures this frequency $\gamma_e$ is the number $2 \times 10^{11}$ multiplied by the density of the air ($\rho$) compared with sea-level density ($\rho_o$), or $\gamma_e = 2 \times 10^{11} \rho/\rho_o$. At altitudes above 30 kilometers, where an area-defense system will have to make most of its interceptions, the density of air is less than .01 of the density at sea level. Under these conditions, the electron collision frequency $\gamma_e$ is less than the value of $\omega = (2\pi \times 3 \times 10^8)$ and therefore can be neglected in the denominator of the equation. Using that equation, we can then specify the number of electrons, $N_e$, needed to attenuate one-meter radar waves by a factor of more than one decibel per kilometer: $N_e > 350\rho_o/\rho$. At an altitude of 30 kilometers, where $\rho_o/\rho$ is about 100, $N_e$ is about $3 \times 10^4$, and at 60 kilometers $N_e$ is still only about $3 \times 10^6$. Thus the electron densities needed for the substantial attenuation of a radar signal are well under the $10^9$ electrons per cubic centimeter required for total reflection. The ionized cloud created by the fireball of a nuclear explosion is typically 10 kilometers thick; if the attenuation is one decibel per kilometer, such a cloud would produce a total attenuation of 10 decibels. This implies a tenfold reduction of the outgoing radar signal and another tenfold reduction of the reflected signal, which amounts to effective blackout.

The temperature of the fireball created by a nuclear explosion in the atmosphere is initially hundreds of thousands of degrees centigrade. It quickly cools by radiation to about 5,000 degrees C. Thereafter cooling is produced primarily by the cold air entrained by the fireball as it rises slowly through the atmosphere, a process that takes several minutes.

When air is heated to 5,000 degrees C., it is strongly ionized. To produce a radar attenuation of one decibel per kilometer at an altitude of 90 kilometers the fireball temperature need be only 3,000 degrees, and at 50 kilometers a temperature of 2,000 degrees will suffice. Ionization may be enhanced by the presence in the fireball of iron, uranium and other metals, which are normally present in the debris of nuclear explosion.

The size of the fireball can easily be estimated. Its diameter is about one kilometer for a one-megaton explosion at sea level. For other altitudes and yields there is a simple scaling law: the fireball diameter is equal to $(Y\rho_o/\rho)^{1/3}$, where $Y$ is the yield in megatons. Thus a fireball one kilometer in diameter can be produced at an altitude of 30 kilometers (where $\rho_o/\rho = 100$) by an explosion of only 10 kilotons. At an altitude of 50 kilometers (where $\rho_o/\rho = 1,000$), a one-megaton explosion will produce a fireball 10 kilometers in diameter. At still higher altitudes matters become complicated because the density of the atmosphere falls off so sharply and the mechanism of heating the atmosphere changes. Nevertheless, fireballs of very large diameter can be expected when megaton weapons are exploded above 100 kilometers. These could well black out areas of the sky measured in thousands of square kilometers.

For explosions at very high altitudes (between 100 and 200 kilometers), other phenomena become significant. Collisions between electrons and air molecules are now unimportant. The condition for blackout is simply that there be more than $10^9$ electrons per cubic centimeter.

At the same time, very little mass of air is available to cool the fireball. If the air is at first fully ionized by the explosion, the air molecules will be dissociated into atoms. The atomic ions combine very slowly with electrons. When the density is low enough, as it is at high altitude, the recombination can take place only by radiation. The radiative recombination constant (call it $C_R$) is about $10^{-12}$ cubic centimeter per second. When the initial electron density is well above $10^9$ per cubic centimeter, the number of electrons remaining after time $t$ is roughly equal to $1/C_R t$. Thus if the initial electron density is $10^{12}$ per cubic centimeter, the density will remain above $10^9$ for 1,000 seconds, or some 17 minutes. The conclusion is that nuclear explosions at very high altitude can produce long-lasting blackouts over large areas.

The second of the two mechanisms for producing an ionized cloud, the beta rays issuing from the radioactive debris of a nuclear explosion, can be even more effective than the fireball mechanism. If the debris is at high altitude, the beta rays will follow the lines of force in the earth's magnetic field, with about half of the beta rays going immediately down into the atmosphere and the other half traveling out into space before returning earthward. These beta rays have an average energy of about 500,000 electron volts, and when they strike the atmosphere, they ionize air molecules.

Beta rays of average energy penetrate to an altitude of about 60 kilometers; some of the more energetic rays go down to about 50 kilometers. At these levels, then, a high-altitude explosion will give rise to sustained ionization as long as the debris of the explosion stays in the vicinity.

One can show that blackout will occur if $y \times t^{-1.2} > 10^{-2}$, where $t$ is the time after the explosion in seconds and $y$ is the fission yield deposited per unit horizontal area of the debris cloud, measured in tons of TNT equivalent per square kilometer. The factor $t^{-1.2}$ expresses the rate of decay of the radioactive debris. If the attacker wishes to cause a blackout lasting five minutes ($t = 300$), he can achieve it with a debris level $y$ equal to 10 tons of fission yield per square kilometer. This could be attained by spreading one megaton of fission products over a circular area about 400 kilometers in diameter at an altitude of, say, 60 kilometers. Very little could be seen by an area-defense radar attempting to look out from under such a blackout disk. Whether or not such a disk could actually be produced is another question. Terminal defense would not, of course, be greatly disturbed by a beta ray blackout.

The foregoing discussion has concentrated mainly on the penetration aids that can be devised against an area-defense system. By this we do not mean to suggest that a terminal-defense system can be effective, and we certainly do not wish to imply that we favor the development and deployment of such a system.

Terminal defense has a vulnerability all its own. Since it defends only a small area, it can easily be bypassed. Suppose that the 20 largest American cities were provided with terminal defense. It would be easy for an enemy to attack the 21st largest city and as many other undefended cities as he chose. Although the population per target would be less than if the largest cities were attacked, casualties would still be heavy. Alternatively, the offense could concentrate on just a few of the 20 largest cities and exhaust their supply of antimissile missiles, which could readily be done by the use of multiple warheads even without decoys.

It was pointed out by Charles M. Herzfeld in *The Bulletin of the Atomic Scientists* a few years ago that a judicious employment of ABM defenses could equalize the risks of living in cities of various sizes. Suppose New York, with a population of about 10 million, were defended well enough to require 50 enemy warheads to penetrate the defenses, plus a few more to destroy the city. If cities of 200,000 inhabitants were left undefended, it would be equally "attractive" for an enemy to attack New York and penetrate its defenses as to attack an undefended city.

Even if such a "logical" pattern of ABM defense were to be seriously proposed, it is hard to believe that people in the undefended cities would accept their statistical security. To satisfy everyone would require a terminal system of enormous extent. The highest cost estimate made in public discussions, $50 billion, cannot be far wrong.

Although such a massive system would afford some protection against the U.S.S.R.'s present armament, it is virtually certain that the Russians would react to the deployment of the system. It would be easy for them to increase the number of their offensive warheads and thereby raise the level of expected damage back to the one now estimated. In his recent forecast of defense needs for the next five years, Secretary McNamara estimated the relative cost of ABM defenses and the cost of countermeasures that the offense can take. He finds invariably that the offense, by spending considerably less money than the defense, can restore casualties and destruction to the original level before defenses were installed. Since the offense is likely to be "conservative," it is our belief that the actual casualty figures in a nuclear exchange, after both sides had deployed ABM systems and simultaneously increased offensive forces, would be worse than these estimates suggest.

Any such massive escalation of offensive and defensive armaments could hardly be accomplished in a democracy without strong social and psychological effects. The nation would think more of war, prepare more for war, hate the potential enemy and thereby make war more likely. The policy of both the U.S. and the U.S.S.R. in the past decade has been to reduce tensions to provide more understanding, and to devise weapon systems that make war less likely. It seems to us that this should remain our policy.

# Meaningless Superiority

For many years politicians of West and East appeared to share our convictions. Both sides pursued very cautious foreign policies, careful to avoid any conflict which might possibly lead to nuclear war. Now, unfortunately, many politicians and others speak of superiority in nuclear weapons and of winning a nuclear war. As Henry Kissinger clearly recognized when he was secretary of state, nuclear superiority is a meaningless concept. There can be no victor in a nuclear war.

Nuclear war would be a complete catastrophe for all participants. The Pentagon estimated some years ago that there might be 100 million dead in the United States, 100 million in the Soviet Union, and a similar number in Western Europe. Who can claim victory in such a war? What life would there be for the survivors? Transportation would probably be thoroughly destroyed. There would be not enough food in the cities, while crops rot in the field for want of fuel to harvest them, and medical services would be interrupted everywhere.

We would be lucky, in the circumstances, if any social structure were left, if at first there were any government at all. It is utopian to think that we could save our system of government and economics or our civil liberties after a nuclear war. All we cherish and are willing to fight for would have been destroyed. The government that would emerge would surely be a dictatorship.

Therefore, what counts in the nuclear age is security against nuclear attack: a reasonably invulnerable second-strike force. Only this, not superiority in number of missiles or warheads or megatons, can give us any security in the age of nuclear weapons. We have such a second-strike force in our missile submarines. No effective antisubmarine weapons exist, and none are likely to be developed in the foreseeable future. Even if such weapons were developed, it would be well nigh impossible to find and destroy all our 41 missile submarines simultaneously. And we must remember that even a few of them would be enough to inflict "unacceptable damage."

In addition, the United States has bombers, many of which will soon be equipped with cruise missiles which will greatly enhance their ability to penetrate enemy territory.

If land-based missiles are becoming vulnerable, why not draw the logical conclusion and give them up? It seems to me futile to try to preserve ground-based missiles by desperate means like the MX, even more futile to deploy ground-based missiles in the still more vulnerable, small territory of Western Europe. We are told that the United States cannot negotiate on arms control until our land-based missiles become invulnerable. Is it possible to achieve this at all? And do we then have to wait until the Soviet missiles have also become invulnerable?

Turning from strategic to tactical weapons, it seems to me a fallacy to believe the use of nuclear weapons can be confined to tactical ones. Once war has escalated to this level, neither side is likely to give up. And escalation to a full-scale strategic, intercontinental war is likely. Tactical nuclear weapons are a very dangerous way to try to stop a Soviet invasion of Western Europe, especially if Russia has used only so-called conventional weapons.

I consider it very unlikely that Russia will launch a major war against Western Europe. What use would a destroyed Europe be to them? But suppose Russia were to attack Western Europe some time in the future? The only way to save Europe would be to counter such an invasion with conventional weapons.

We have developed "smart bombs" and other weapons of high accuracy. Let us develop these further and plan to use them without nuclear warheads. Once we feel sufficiently confident about them, we may at last be able to declare that we will not be the first to use nuclear weapons, and probably the Russians would join in such a declaration.

The most sophisticated and precise weapons will be useless unless the armed forces, in all services, know how to use and maintain them and have the morale to persevere.

Increasing the number of our nuclear armaments is futile because it leads to further increases by the Soviets, as already announced by the Soviet Chief of Staff. Clearly, in the nuclear age, the physical survival of the major countries is irrevocably entwined.

The only way to enhance our security, and that of our allies, is to negotiate arms-control agreements, including substantial arms reductions. Proposals for such reductions must be equitable, otherwise the other side will reject them out of hand.

Negotiations on arms control must not be linked to "good behavior" of the Soviets in international affairs. We Americans should have learned in Vietnam that we are not the policeman of the world. In fact, in times of visible conflicts of interest between Russia and us, control of nuclear armaments is even more necessary than in times of detente. Proper nuclear arms control and reduction is not a unilateral advantage, it benefits both sides equally. We must have reduction of nuclear armaments, not an arms race which leaves both sides in still more deadly peril.

# We Are Not Inferior to the Soviets

Hawks and doves agree on at least one thing: that nuclear armaments are excessive and must be reduced. But the present administration has downgraded the importance of serious arms-control negotiations by giving first priority to adding many new weapons to the U.S. and NATO arsenals: the B-1 bomber, MX missile, and Trident II missile in this country; and the Pershing II missile and ground-based cruise missiles in Europe. Without these, President Reagan and former Secretary Haig have insisted, the United States would be caught in a position of permanent inferiority. We are told that there is a serious "window of vulnerability" in our forces.

I claim that our strategic nuclear forces are not inferior to those of the Soviets. Let us look at the actual numbers.

Table 1 compares the numbers of delivery vehicles of various kinds possessed by the U.S. and the U.S.S.R. The Soviets have more intercontinental ballistic missiles (ICBMs) and more submarine-launched ballistic missiles (SLBMs), but fewer bombers. In total, the Soviets have more delivery vehicles than the U.S., plus Britain and France, but the difference is not significant.

### TABLE 1. Delivery Vehicles

|  | U.S. | U.S.S.R. |
|---|---|---|
| ICBMs (intercontinental ballistic missiles) | 1,050 | 1,400 |
| SLBMs (submarine-launched ballistic missiles) | 630[a] | 950 |
| Bombers | 350 | 140 |
| Total | 2,030[b] | 2,490 |

a. Polaris: 112, Poseidon: 352, Trident II: 168.
b. Plus 144 British and French delivery vehicles.

In "equivalent megatons," the Soviet nuclear force is about twice that of the U.S. (see Table 2; "equivalent megaton" is defined there). The Soviets have put larger-yield weapons on their missiles, an advantage that is cancelled by the lower accuracy of their missiles.

#### TABLE 2. Equivalent Megatons[a]

|  | U.S. | U.S.S.R. |
|---|---|---|
| ICBMs | 1,300 | 5,900 |
| SLBMs | 800 | 1,200 |
| Bombers (present) | 3,500 | 900 |
| Present total | 5,600 | 8,000 |
| Total (according to Nuclear War)[b] | 4,100 | 7,100 |
| Bombers (after cruise missiles deployed) | 2,700 | — |
| Revised total | 4,800 | |
| 1985 total (according to Nuclear War)[b] | 4,200 | 9,200 |

a. A weapon's yield in megatons raised to the two-thirds power gives its yield in "equivalent megatons." Thus, a 2-megaton bomb contributes 1.59 equivalent megatons to the arsenal and a .5-megaton bomb contributes .63 equivalent megatons. Comparing equivalent megatons is, approximately, comparing areas that can be destroyed. (Equivalent megatons is not the right measure for comparing fallout.)

b. *Nuclear War: What's In It For You?*, a book prepared by Ground Zero (Simon and Schuster, New York, 1982).

The most important comparative measure of strength is number of warheads. By this measure (Table 3), the U.S. is somewhat ahead and is expected to remain ahead.

#### TABLE 3. Warheads

|  | U.S. | U.S.S.R. |
|---|---|---|
| ICBMs | 2,150–2,250 | 5,500–6,400 |
| SLBMs | 4,750 | 1,750–1,900 |
| Bombers (present) | 2,500–3,500 | 280–550 |
| Present total | 9,400–10,500 | 7,530–8,850 |
| Bombers (after cruise missiles deployed) | 4,700 | |
| 1985 estimated total | ~12,000 | ~10,000 |

A single warhead can destroy nearly any target, industrial or military, other than hardened silos. The smallest nuclear weapon in our arsenal is 40 kilotons, about three times the explosive energy of the Hiroshima bomb. (Megatons are only important as weapons of pure terror, to threaten populations.)

Soviet accuracy has improved in recent tests. In a few years, the Soviets could, in principle, eliminate much of our ICBM (Minuteman) force in a first strike. This is the perceived window of vulnerability. (At present, they could destroy perhaps half of the Minuteman force.)

*But* such an attack would have no possible military advantage for the U.S.S.R. It would leave this country with 75 percent of its weapons available. In fact, it was foreseen a long time ago—in the 1950s—that the time would come when ICBMs would be vulnerable. For this reason, the U.S. nuclear force was diversified, and has remained so. As shown in Table 4, ICBMs will soon account for only one-fourth of U.S. warheads.

#### TABLE 4. Warhead Percentage Distribution
(Deliverable warheads[a], after introduction of 3,000 cruise missiles by the U.S.)

|                | U.S.  | U.S.S.R. |
|----------------|-------|----------|
| ICBMs          | 25%   | 77%      |
| SLBMs          | 53%   | 23%      |
| Cruise Missiles | 22%  | 0        |
| Total          | 100%  | 100%     |
| Total number   | 9,000 | 7,800    |

a. For this table, it is assumed that all ICBM and SLBM warheads are "deliverable," that two-thirds of the B-52-launched cruise missiles are deliverable, and that a negligible percentage of weapons carried by Soviet bombers are deliverable.

Of the Polaris submarines deployed in the period 1959–1967, thirty-nine are currently in service. They are essentially invulnerable. Submarines carry half of the U.S. warheads. Destruction of the Minuteman force would not be disabling. Such a first strike by the U.S.S.R. would be madness.

In addition to submarines, we have bombers plus cruise missiles. Three-thousand cruise missiles are to be installed on 150 B-52s. There is no effective air defense against cruise missiles.

Table 5 shows the present high accuracy of Soviet missiles and even higher accuracy of U.S. missiles. These accuracies will be improved in the future.

### TABLE 5. Missile Accuracy[a]

|  | U.S. | U.S.S.R. |
|---|---|---|
| ICBMs (present) | 600–1,000 ft | 1,000–1,500 ft |
| ICBMs (future) | ~300 ft | ? |
| SLBMs | 1,500 ft | 3,000 ft |
|  |  | 5,000 ft |
| Cruise missiles | 300 ft | (land-based) |
|  |  | 50 ft |
|  |  | (ship-to-ship) |
| Future missiles: terrain-scan cruise missiles and perhaps eventually ICBMs and SLBMs | 50–100 ft | ? |

a. The numbers are taken from *Nuclear War* (see Table 2).

Extremely high accuracy is needed only for hard targets (silos). For most military targets, SLBMs are sufficiently accurate. Cruise missiles on bombers can have any accuracy we may want, and bombers can be on alert in time of tension, or can take off from widely dispersed airfields on warning.

Therefore, our strategic forces as a whole will not become vulnerable. Our Minuteman missiles may become vulnerable, but they are only a small part of our forces. *The window of vulnerability does not exist.*

If at all, such a window may exist for the Soviets. They have put most of their strength (and warheads) on ICBMs. If ICBMs become vulnerable, the Soviets are much more vulnerable than we are. *Because of the distribution and the greater invulnerability of our forces, we are certainly not inferior, but are in fact superior to the Soviets in strategic weapons.*

The same opinion is held by military men. The chief of staff of each of the services was asked in 1981 whether he would trade his service—its weapons, personnel, missions, entire range of capabilities, strengths, and weaknesses—for its Soviet counterpart service. Each of the generals and the admiral said he would not make such a trade.

Most important: Comparison of numbers is meaningless for nuclear weapons, beyond a certain minimum number. Both the United States and the Soviet Union have vast overkill capability. If you wish to destroy the other country's important military installations, other than its ICBMs, a few hundred warheads are enough; to destroy the more important industrial plants, another few hundred, and in neither case do they need to have extreme accuracy. There is no justification for the many thousands of

94

ARMS CONTROL

nuclear warheads that each of the two superpowers possess. Superiority or inferiority, at present levels, has no meaning.

We do not need the B-1 bomber. Cruise missiles launched from B-52 bombers can penetrate defenses to reach targets better, cheaper, and more reliably. If we need bombers in a small peripheral war, the elaborate (and costly) electronics on the B-1 are not needed.

For the same reasons, we do not need the follow-on to the B-1, the Stealth bomber.

Nor do we need the MX missile. President Reagan deserves credit for eliminating the "shell game" basing mode for the MX, but I cannot see *any* basing of ICBMs on land that will remain safe.

We probably do need further Trident submarines to replace some of the aging Polaris submarines. We may want the Trident II missile with its super-high accuracy, but this could be negotiable in an arms-control treaty.

There is a true window of vulnerability that is now wide open. It is that all of us, in the Soviet Union, the United States, and Western Europe, are constantly exposed to the danger of a nuclear war that might kill hundreds of millions of people, and would destroy civilization. This danger is heightened by statements like "We can survive a nuclear war" or even "We can win a nuclear war." No country can win a nuclear war; there are only losers. Chairman Brezhnev has said this clearly: "It is a dangerous madness to try to defeat each other in the arms race and to count on victory in nuclear war. I shall add that only he who has decided to commit suicide can start a nuclear war in the hope of emerging a victor from it. No matter what the attacker might possess, no matter what method of unleashing nuclear war he chooses, he will not attain his aims. Retribution will ensue ineluctably." I am quoting Brezhnev to counter the claim by some influential people in the U.S. government that the Russians consider nuclear war winnable.

The main imperative is to see to it that there never be a nuclear war. This must have priority over comparisons of the strength of nuclear forces, which are meaningless anyway.

Beyond this, we must greatly reduce the level of the nuclear armaments of both superpowers. This is the meaning of arms control.

There have been two arms-control agreements, SALT I and SALT II. SALT I is in force. SALT II was negotiated, painfully, in seven years and three administrations, and was signed by President Carter and Chairman Brezhnev, but was never ratified by the U.S. Senate. This is a pity, because it was a good, carefully balanced treaty and to our advantage—the Soviets would now have fewer missiles than they actually have if the treaty were in force. Our government has withdrawn the treaty from consideration by the Senate, but, for the present, the planned arms buildup remains within the limits set by SALT II.

The trouble with the SALT agreements is not that they were unbalanced in favor of the Soviets, but that our military (and presumably theirs as well) always demanded some buildup of forces as a price for agreeing to the SALT treaties. The price for SALT II was paid (the MX), but the treaty was not ratified. In this way, strategic forces have constantly increased in spite of treaties.

There are many proposals to correct this situation. The best known one is the Nuclear Weapons Freeze. There is a popular movement for the Freeze; there has been a popular vote in favor of it in New Hampshire; there is an Initiative on the ballot in California; and an opinion poll has shown that 72 percent of the American people are in favor of it. There are two Freeze resolutions in the Senate, one by Senators Kennedy and Hatfield, the other by Senators Jackson and Warner.

The Kennedy-Hatfield Freeze seems simple and straightforward, but is not well defined. Taken literally, it might mean that no change of nuclear weapons is permitted at all. This would prohibit, for example, our installation of cruise missiles on bombers. This would be a highly undesirable limitation because cruise missiles clearly constitute a second-strike force. Both sides should be permitted to change their weapons arsenal to make it more survivable, as long as the total number and yield of nuclear weapons are not increased. For instance, the Soviets should be permitted and even encouraged to change more of their weapons from ICBMs to submarines. This is not only to their advantage but also to ours, because it would remove their incentive to use their ICBMs quickly, in one big first strike, lest they be destroyed.

The Jackson-Warner "Freeze" is far too permissive. It would allow the United States to build up all the additional weapons in the Reagan program before there is a freeze, without paying attention to what the Soviets may do in the meantime. This is no freeze at all.

There is an important idea in the Freeze, namely that we should do something on our own initiative to stop the arms race. Negotiations with the Soviets are slow and frustrating at best, and something must be done *now* to keep the arms race under control. At the same time, we cannot afford unilateral disarmament.

A very good idea has been proposed by George Kennan, and modified by Jeremy Stone and Robert Bacher. Kennan proposed that both the U.S. and the U.S.S.R. cut their forces to half of the present level. This could be done without diminishing the security of either power. However, it is unlikely to be politically feasible. The modification by Stone and Bacher proposes that we cut our forces unilaterally by 5 percent and challenge the Soviets to do the same. Through satellite reconnaissance, we can discover whether they are actually doing this. If they are, we cut again 5 percent next year, and we continue this decrease year after year. In the meantime, we negotiate very seriously, with the aim of achieving Kennan's 50-percent cut in a logical and verifiable manner.

Controlling nuclear armaments is not done as a favor to the Soviets. It is done to reduce a mortal threat to America, and to the whole world. Many secretaries of defense have recognized that we are more secure with arms control than without it, and so have many of the chiefs of staff. Arms control negotiations must be undertaken no matter what the state of our relations with the Soviets are on other concerns. In fact, in times of crisis, it becomes even more important to have a good arms control agreement in force. I am happy that the arms control problem has, in fact, been decoupled from the general foreign policy of the United States toward the Soviet Union. Our delegates are meeting in Geneva to discuss the control of "theater nuclear weapons" in Europe.

Nuclear weapons in Europe are a special problem. For decades, the United States and our allies in NATO have considered nuclear weapons an effective deterrent against a hypothetical Soviet attack on Western Europe using conventional weapons. This was reasonable in the early years after World War II, when the Soviets had overwhelming superiority in conventional weapons over an exhausted Western Europe, and conversely, the United States had, first a monopoly, and later overwhelming superiority, in nuclear weapons over the Soviets.

In the long period since World War II, the situation has changed. There are two great dangers in using nuclear weapons in a hypothetical war in Europe. The first is escalation: The use of any nuclear weapon, however small, on the battlefield may lead the opponent to use a bigger one, and this may continue until megaton weapons are used. The Soviets have emphasized this concern many times: They consider any use by us of nuclear weapons in Europe as having crossed the nuclear threshold, and as a justification for them to use any type of nuclear weapons anywhere. Battlefield use of nuclear weapons thus involves the enormous risk of all-out nuclear war. But even if this situation did not happen, the use of nuclear weapons in Europe would very likely destroy much of Europe (the second great danger), because of the very high population density, and because of the enormous destructive power of nuclear weapons. Thus, in an effort to save Europe, we would destroy it. Europeans are acutely aware of this dilemma.

In view of this situation, four highly respected former government officials—McGeorge Bundy, George Kennan, Robert McNamara, and Gerard Smith—have introduced a new proposal. They propose that we consider a declaration that we will never be the first to use nuclear weapons. The Soviets have previously challenged us to make such a declaration. Obviously, a mutual declaration would not ensure that nuclear weapons will never be used, and would not permit us to abandon such weapons completely. But it would greatly reduce the probability that they will be used.

Former Secretary Haig was quick to object that this would leave Europe open to an invasion with conventional weapons from the East, and that it would be extremely costly to bring NATO conventional weapons up to equality with those of the Warsaw Pact. This is very likely not correct.

Table 6 compares the conventional forces of East and West in Europe, along with other measures of relative strength. The Warsaw Pact nations do indeed have superiority in tanks, in the ratio of 3 to 2. But NATO has an enormous number of "smart" antitank weapons, more than 10 for each Warsaw Pact tank in Europe. These are more effective and reliable than neutron bombs. Generally, the forces of East and West are well matched. Presumably, we would have to increase our conventional armaments in some areas, but I am told on good authority that this could be done for about $10 billion, a quarter of the price of the B-1 bomber program alone. We may want to postpone the actual declaration of "No First Use" until these improvements are in place. But I believe that Robert McNamara and his associates have advocated a most important idea.

The arms race has to be stopped.

### TABLE 6. Conventional Forces in Europe and Related Measures of Relative Strength[a]

|  | NATO | Warsaw Pact |
| --- | --- | --- |
| Tanks (total) | 28,000 | 63,000 |
| Tanks in Europe | 17,000 | 26,000 |
| Antitank missiles | 300,000 + | ? |
| Combat aircraft (total) | 10,500 | 10,800 |
| Land-based aircraft in Europe | 4,280 | 4,950 |
| Naval aircraft | 1,150 | 770 |
| Ground forces in Europe | 2,123,000 | 1,669,000 |
| Military manpower | 4,900,000 | 4,800,000 |
| Annual military spending | $241 billion | $202 billion |

a. Source: "Soviet Military Power: Questions and Answers," *Defense Monitor*, Vol. 11, No. 1 (1982).

The claim that the United States is inferior to the Soviet Union in strategic nuclear armaments is wrong. The claim that the conventional forces of NATO are hopelessly inferior to those of the Warsaw Pact nations is also wrong. These claims needlessly fuel the arms race.

I have been somewhat critical of the proposed Nuclear Weapons Freeze. But it has the right spirit. The people devoted to ending the arms race must not fight each other, but must stand together. Only by ending the arms race and then decreasing nuclear armaments can the United States and the world find real security.

# The Five-Year War Plan

## WITH KURT GOTTFRIED

The Soviet Union has always been blessed with five-year plans. Thanks to the Defense Department, we are to have our own in the "Defense Guidance" that is to form the basis for Pentagon budget requests for the next five fiscal years. This comprehensive plan which envisions a possibly "protracted" nuclear war, global conventional war and "space-based" fighting with antisatellite weapons, must be examined with great care. It comes close to a declaration of war on the Soviet Union and contradicts and may destroy President Reagan's initiatives toward nuclear arms control.

The plan reveals a nonchalance toward nuclear war, an inability to distinguish real dangers from farfetched nightmares, an unwillingness to learn that many of our technological breakthroughs have returned to haunt us — that the arms race is an increasingly dangerous treadmill.

America, always the pacesetter in weapons development, must learn to ask whether innovation ultimately benefits the Russians more than us. Take, for example, the multiple warhead missile, or MIRV, which we introduced even though the Arms Control and Disarmament Agency warned that eventually it would be advantageous to the Russians because their larger missiles could accommodate more warheads. The exaggerated concern that our land-based missiles are vulnerable would not exist had we negotiated a ban on MIRVs.

The development of antisatellite weapons, proposed by the Pentagon, would jeopardize our security. Satellites are far more important to us than to the Russians. The Western alliance is an open book. *Aviation Week* publishes important military data regularly. American officials and Presidential candidates divulge highly sensitive information when seeking funds or votes.

Most of our intelligence about the Soviet military comes from spy satellites. They provide the only effective verification of compliance with the SALT treaties. Before satellites, our assessment of Soviet strength was often wide of the mark: The 1960 missile "gap" dissolved when our first satellites showed the Russians were far behind. Satellites also preclude a surprise attack by conventional forces: Soviet preparations for the invasion of Czechoslovakia in 1968 were quickly discovered by satellites.

Should conventional or nuclear hostilities break out between the super-powers, satellites would be crucial. Feverish negotiations surely would accompany the calamity, and satellites would warn us of deception. Enemy forces' movements could be watched. If satellites were destroyed as the first act of war, both countries would be blinded Goliaths bent on murder, but America would be more severely handicapped.

Moscow has sporadically tested a very crude antisatellite system for about 10 years. Its foolishness should not impel us to lose our heads. It makes it even more urgent to prevent deployment of antisatellite weapons. This cannot be done by building and testing such weapons, for that can only provide a fleeting advantage. As with MIRVs, our security and the Russians' would be best served by a negotiated agreement forbidding all testing of such weapons.

The five-year plan displays a profoundly disturbing attitude toward nuclear weapons. In contrast to the statesmanlike posture that President Reagan and Secretary of State Alexander M. Haig, Jr. have now adopted, it bristles with nuclear saber-rattling: limited winnable nuclear warfare, nuclear weapons in space, antiballistic missile defenses. There are glaring inconsistencies: Our nuclear weapons would "render ineffective the total Soviet . . . power structure," while our "plan" assumes Mr. Reagan and his staff could control a nuclear exchange. That some of these steps could violate treaties (nuclear weapons in space, the ABM pact) appears not to matter to the Pentagon.

Indeed, the plan proposes to address all threats—real and imagined—by raising the ante. It refuses to recognize that our worst nightmares can be laid to rest only by constraints on technology. Verifiable treaties toward that end have been proposed but are opposed by the Pentagon. These trea-ties would forbid all underground nuclear tests and severely limit missile test firing.

If the Pentagon plan becomes government policy, the arms race will quicken. The plan's notion that we could win such a race reveals an igno-rance of post-1945 history.

The plan rightly opposes the transfer of sensitive technology to the Rus-sians. Does it question the zeal with which the Pentagon sells sophisticated weapons to unstable third-world countries? It calls for stupendous growth of the armed forces after 1988. Do the planners believe that the electorate would approve the militarization of our economy that this would require? The race proposed by the Pentagon would make the threat to civilization even more unacceptable than it is today. Must we have a plan that will do for our national security what a Soviet five-year plan does for agriculture?

# Debate: Elusive Security

## WITH KURT GOTTFRIED
## RESPONSE BY MALCOLM WALLOP

The Reagan administration's attitude toward nuclear weapons and nuclear war has profoundly alarmed millions of Americans. The unexpected explosion of public support for a nuclear freeze sends a message that is loud and clear: "Stop the arms race, we want to get off!" There is also deep frustration in the land about what can be done when the government in power is as inflexible as this one seems to be.

But something can be done—by Congress. Of course the House and Senate cannot dictate national security policy to the executive branch. But with legislation, resolutions and the right kind of hearings, Congress can transform the political environment in which national security policy is made.

That policy is now based on what we may call the Weinberger Doctrine:

- The United States must have the ability not just to wage nuclear wars but to "prevail" in such conflicts.
- By engaging in an arms race, the United States can force the Soviet Union to adopt policies more to our liking.

This doctrine was unveiled when the Five-Year Defense Guidance Plan, signed by Defense Secretary Caspar Weinberger, was leaked to the press in May 1982. It has never been disavowed by the Reagan administration.

That document states that our nuclear forces must be able to "decapitate" the Soviet leadership, while we control the climb up the "escalation ladder." The plan expects us to "prevail" and to attain an end of hostilities on "terms favorable to the U.S.," though it neglects to explain with whom these terms would be negotiated.

The Reagan administration differs from all its predecessors in one important respect: The formulation and execution of defense policy is now in the hands of civilians who believe that this country really could "prevail" in a nuclear war. Weinberger's five-year plan heralds the triumph of a school of civilian strategists whose theories tell us as much about national security as astrology does about the nature of the heavenly bodies. No wonder the Joint Chiefs of Staff have had to become the restraining force in the Pentagon.

The Weinberger Doctrine is firmly based on misconceptions, some military, others political.

On the military side, it fails to recognize that any nuclear war—no matter how limited—could escalate to a strategic nuclear war. One can argue forever whether this escalation probability is 10 percent, or 50 percent, or 90 percent. But it is certainly not zero.

Once a strategic exchange begins, escalation to an all-out holocaust becomes very possible. Both human and technical factors point in that direction. In particular, the intricate systems that both superpowers use to control their nuclear weapons, and to watch their opponent, are highly vulnerable to attack. Hence the use of nuclear weapons to achieve any military or political objective whatever entails unacceptable risks. Nuclear weapons can only serve as a deterrent.

The political facet of the Weinberger Doctrine is even more remarkable. While American voters will never accept the arms race that it envisages, the doctrine will provoke a Soviet response that could last for a generation, because the dictatorship of the proletariat is largely immune to the desires of proletarians. The notion that the White House could compete with the Kremlin in extracting the sacrifices demanded by a more intense arms race is nonsense.

One might choose simply to await the inevitable failure of the Weinberger Doctrine. But while our leaders chase the mirage of nuclear superiority, opportunities that will never return are lost.

Propaganda notwithstanding, Soviet positions in three vital areas of arms control deserve serious consideration: on nuclear proliferation, weapons in space, and medium-range missiles in Europe. All have been rejected or ignored by the Reagan administration. All deserve the scrutiny of Congress.

The opportunities that we could soon lose are best appreciated by placing them in a historical setting.

After Hiroshima, the United States tried in vain to achieve international control over nuclear weapons. In retrospect, it is obvious that Stalin was far too suspicious and power-hungry to allow America to be the only nation to possess such a revolutionary military technology.

In 1958 President Eisenhower announced that we were willing to negotiate a Comprehensive Test Ban Treaty (CTBT) outlawing all nuclear tests. This imaginative proposal went down with the U.S. spy plane shot down over Russia. No president since then has had both the wisdom and prestige required for so far-reaching a step.

By the time the 1963 treaty banning atmospheric tests was signed, the arms racers had gutted the Eisenhower concept: Testing just went underground. Had the comprehensive ban been adopted in 1960, the most destabilizing weapons in both arsenals—the multiple warhead missiles (MIRV)—would now be more primitive and less dangerous.

In 1974 President Nixon signed the Threshold Test Ban Treaty prohibiting underground tests larger than 150 kilotons—some 10 times the Hiroshima bomb. It has not been ratified. President Carter made progress toward a comprehensive test ban. The Russians agreed to a network of seismic detectors on Soviet soil, and to some on-site inspection. But Carter did not have the political strength to capitalize on this. In July 1982 the Reagan administration announced that it would be the first U.S. government to abandon all negotiations towards a CTBT.

The evolution of the strategic balance is also instructive, because in advocating the Weinberger Doctrine the administration often rewrites history.

On November 22, 1982, President Reagan, while pointing to a chart, said to the nation, "Believe it or not, we froze our numbers in 1965 and have deployed no additional missiles since then." He neglected to say that since 1965, because of MIRV, the number of warheads carried by our submarines and land-based missiles has grown from 1,700 to 6,200.

Reagan also said, "Many of our bombers are now older than the pilots who fly them," but he did not point out that, at this moment, they are being equipped with 3,000 incredibly accurate nuclear cruise missiles that can be fired from outside Soviet air defenses. Therefore no new high-performance plane, such as the B-1, is needed to bomb the Soviet Union.

In brief, today our strategic weapons are far more numerous, survivable, versatile, accurate and reliable than in 1965.

Since 1972 the Soviet Union has quadrupled its strategic warheads and greatly strengthened its European missile force. But we should remember that we have also made dramatic moves.

In 1960–1965, the advent of Polaris submarines and Minuteman ICBMs produced a massive increase in the potency of our forces. That's why the administration uses 1965 as the base year in promoting its buildup; it does not mention that we then enjoyed a 10-to-1 advantage in warheads. And the recent Soviet buildup was certainly triggered by our earlier MIRV deployment.

What can Congress do to stem this tide?

Consider, first, that the administration has abandoned the test ban negotiations, and has even indicated a desire to renegotiate the Threshold Test Ban Treaty, because it claims we cannot verify Soviet adherence to the 150-kiloton limit. This contradicts the opinion of virtually all leading geophysicists. In any event, if that is really the president's concern, he should ask Congress to ratify the Threshold Treaty, whereupon both parties must hold joint test explosions to check the calibration of their seismic detectors.

But the current 150-kiloton limit is a red herring. All experts agree that underground explosions as small as two kilotons can be detected and distinguished—with high confidence—from earthquakes. A test ban with an exceedingly low threshold is verifiable.

Unfortunately, by now a comprehensive ban would have little impact on the superpowers' arms race. Soviet and U.S. warheads are already approaching the ultimate level of perfection. But a comprehensive ban would be a severe hurdle for nations trying to join the nuclear club.

Congressional committees should hold hearings on the detection of underground explosions and on the impact of a test ban on proliferation, with a view toward a congressional resolution calling on the administration to negotiate a Comprehensive Test Ban Treaty.

Next, note that the Weinberger Doctrine calls for "a strategic warfighting antisatellite system," and states that "we must insure that treaties do not foreclose opportunities to develop these capabilities." So spending on the militarization of space is to grow faster than the defense budget as a whole.

Satellites are one of the few godsends to come our way since 1945. Their disappearance from the sky would dramatically increase the danger of nuclear war. They are essential to surveillance, communications and control.

Because the United States is an open society, and our forces are flung across the globe, satellites are more important to us than to the Russians. Should we stumble into a nuclear war, the negotiations that should follow (assuming the adversary had not been "decapitated") would be entirely dependent on satellites for communication and surveillance. All previous presidents, beginning with Eisenhower, have concluded that it would be to our advantage if no nation had any antisatellite capability.

During the Carter administration there were unsuccessful negotiations aimed at this goal. In August 1981 the Soviets proposed at the United

Nations to prohibit the orbiting of weapons of any kind while permitting operation of the shuttle. This administration has not explored whether the Soviets would extend this idea to a formal ban on the testing of antisatellite weapons.

As Senator Larry Pressler, chairman of the Senate subcommittee on arms control, has stressed, it is essential to prevent the testing of the new generation of antisatellite weapons. For that reason he has proposed a mutual freeze on flight tests of such systems while negotiations take place. There is an urgent need for congressional hearings to explore the Pressler proposal and the administration's heedless militarization of space.

Third, recall that in December 1982 the new Soviet leader, Yuri Andropov, offered to halve the number of SS-20 ballistic missiles in Europe if we refrained from installing the new Pershing II and cruise missiles that NATO decided in 1979 to deploy in Europe to counter those SS-20s. Andropov contends that this would give the Soviets the same size force as the West, a conclusion he reaches by counting British and French missiles.

He has been curtly rejected by Washington, Paris and London, and with somewhat less vigor by Bonn. The Reagan administration continues to insist on its "zero option," as though the Soviets could be expected to dismantle all of their rockets for a promise that we will not deploy ours.

A tentative understanding reached last summer by Ambassador Paul Nitze, the Reagan administration's negotiator in Geneva, and his Soviet opposite number—though later rejected by both governments—points toward a solution that would be more desirable than either Andropov's offer or the zero option. That would be a sharp reduction of Soviet missiles in Europe in return for American deployment of some cruise missiles, but no Pershing II ballistic missiles.

The negiations must be pursued vigorously. The placing of U.S. missiles on European soil would be highly divisive. Though we doubt the Russians are that shrewd, one must wonder whether their reckless SS-20 buildup was actually intended to provoke us into a response that would tear NATO apart.

Congressional committees should explore the long-term consequences of deploying hundreds of new missiles in Europe, asking how such deployments could possibly enhance Western security.

If the NATO nations really want to increase their security, they would do better to divert funds from nuclear weapons that could destroy them to conventional weapons that can defend them. Congress should weigh every funding request for nuclear weaponry against what the same money would buy in combat supplies, fortifications, prepositioned equipment and the other humdrum needs of a sound and credible defense.

It is true that it is difficult for Congress to provide leadership in national security affairs, but it is also true that the administration is headed in a disastrous direction. The 1982 elections show that Americans do not want to go in that direction, like lemmings to the sea.

Congress can, and should, provide an alternative. It can withhold funds from military programs that undermine our security. It can pass resolutions in support of the freeze and other critical arms control measures. And it can ratify SALT II and the Threshold Test Ban as executive agreements.

---

## MALCOLM WALLOP RESPONDS

C ritics of the MX missile make some valid points:
MX is the brainchild of wrong-headed thinking about nuclear war. It is born of the utopian thought that we can deter the Soviets by adding megatonage to our arsenal without figuring out what we want to do with it. The MX is too big and has too many warheads, and it probably has to be based in stationary silos that cannot be hidden or fully protected.

Nonetheless, it is critical that we press ahead to build and deploy some MXs. This is the only weapon close to production that could really save American lives in the event of nuclear war.

The Congress must stop dawdling with a "maybe yes, maybe no, or maybe later" approach to this missile as the Soviets relentlessly build strategic weapons according to clear military priorities: to protect themselves and defeat the United States by destroying American weapons.

Arms controllers must stop dreaming about making the problem of the Soviet threat magically disappear. The United States does not need another ineffectual arms-control treaty. We need not only the MX but a new generation of smaller, mobile, concealable and invulnerable missiles. We need antimissile missiles and space-based lasers to destroy incoming weapons in any nuclear exchange.

We need hardware based on hard-headed realizations that we must offer our people some prospect of survival if the horrors of nuclear war are visited upon us.

Today, our nation's safety continues to deteriorate. The Soviet SS-18, a more powerful "counterforce" weapon than the MX (meaning that it can kill enemy missiles in their silos), has been deployed for six years now.

Soviet preparations for intercepting American aircraft and missiles grow by the day. The Soviet's first-rate antiaircraft defense is being augmented by state-of-the-art equipment which may be capable of intercepting ballistic missiles as well.

In addition, the Soviets have extensively tested a new antimissile-missile system. For all we know, it may be in clandestine production. The big battle-management radars for a nationwide antimissile defense already exist. Finally, the Soviets are committing vast resources to laser weapons, and, according to a national intelligence estimate quoted by *The New York Times*, are expected to test a space-based laser in the mid-1980s.

Soviet contingency plans and exercises are depressingly realistic. Recent exercises have practiced attacks on American strategic forces on the ground with SS-18s.

The Soviets also practice intercepting whatever American warheads survive and are launched. All of this means that as the months and years pass, as the Soviets become better able to threaten us and protect themselves, the costs of a possible military confrontation fall for the Soviet Union, and rise for the United States.

On our side, the pace of innovation for strategic weapons has slowed to a crawl, while the Pentagon neglects every proposal for protecting the U.S. population against Soviet strategic weapons. Worse, we do not seem to proceed from a coherent design of what we might do, if worse came to worst, to protect ourselves and defeat the enemy.

This is because, since the days of Robert McNamara, the Pentagon has not conceived and presented strategic weapons with the seriousness they deserve. Instead, it has conceived and built them as means of threatening the Soviet Union with a certain level of damage, and therefore as assurances that the Soviets will neither attack nor blackmail us.

This was well enough so long as Soviet strategic forces were inferior to ours. But today, when one-fifth of the Soviet Union's missile force can deliver more explosive power than all our strategic forces, and as Soviet defensive preparations grow, we must think differently. We must ask what we would do if the Soviets threatened—or tried—to take out most of our forces with a small portion of theirs, saving the rest for coercion?

Many would have us believe that these questions will never arise because the Soviet leaders would never find it in their interest to risk the survival of their society by threatening or carrying out such an attack. But what if, as the evidence suggests, the Soviets really mean to take advantage of a situation in which they can threaten us while protecting themselves? Then what good would our weapons do us in a crunch?

Fundamentally, we have to choose between two ways of looking at nuclear weapons, two ways of trying to ensure that we never have to go to war; either we regard them as doomsday machines with which to threaten the Soviet Union with a certain level of damage, or as tools we can use to provide for our own safety if we are ever attacked. The Pentagon has largely chosen the first alternative.

One reason for this is that many officials fear the American people would be too frightened if their leaders talked of weapons as things which might someday be used. The Pentagon thinks that Americans, like so many children, want to be reassured that war won't happen, and that the MX will help make sure it won't. Thus the gimmicky name "Peacekeeper."

This disrespect for the people's good sense is altogether out of place in a democracy. Americans want to be protected, and sense when they are not. The American people do not demand to be sedated by utopian reassurances. Rather, in big things like strategic weapons, as in little ones, they ask "What's it good for?" "What will it do for me?"

During the debate over the MX, the Pentagon should have given the American people the hard truth that this country desperately needs counterforce missiles like the MX because, despite our best efforts, we might someday be attacked. In that case, every American counterforce missile would save countless American lives.

The Pentagon's handling of the MX is a vivid example of its disregard for both strategic and political reality. MX was conceived and pushed with little understanding of what good it might do for the country if and when it ever has to be used. In 1973 the Department of Defense committed itself to the 10-warhead, 200,000 pound MX as a means of counterbalancing the Soviet Union's deployment of heavy missiles. This monster-sized missile is one of the bad results of the strategic arms limitation (SALT) process, which was supposed to cap the arms race by limiting the number of missile silos on both sides.

Arms-control advocates believed that, with the number of launchers limited, the strategic relationship between the U.S. and the Soviet Union would remain stable indefinitely; each could do grievous harm to the other's population, but neither could protect itself by attacking the other side's missiles.

For years, fixed land-based missiles in general, and the MX in particular, were thought to be "survivable"—and SALT was thought viable—

because the national intelligence estimates were saying that the Soviets were not even trying to develop multiple warhead missiles capable of destroying our silos.

They were wrong. By 1978, when the evidence forced the U.S. intelligence community to admit that the Soviets had developed multiple warheads with accuracies sufficient to destroy our missiles in their silos, the Carter administration and the U.S. Air Force should have reevaluated their understanding of Soviet strategy and their commitment both to the MX and to the SALT process.

In my view there is no longer any technical argument against the proposition that the Soviets — and modern technology — will keep our land-based missiles vulnerable so long as they are in known positions.

Some have asserted that the Soviets would be deterred from trying to exploit our vulnerability because of uncertainties about their own technical abilities to conduct a coordinated, accurate attack against our missiles. These assertions should not be taken seriously. Many experts believe that the Soviets long ago solved the technical difficulties involved in such an attack. Moreover, they believe it is dangerous to rely on the hope that something doable will not be done.

U.S. defense planners have failed to draw the logical conclusions from the Soviets' growing ability to carry out a disarming strike, less because of technology than because of their utopian belief that the Soviets are interested in stability, and of course to the bureaucracy's attachment to its own ongoing programs.

Since 1978, we have witnessed a series of attempts to do the impossible: to find a means by which our valuable, relatively immobile MX missile could survive in *known* locations. The American people have been asked to believe that if only the right basing mode could be found, the "ideal" situation of the 1960s would be reestablished, and the balance of terror would remain stable forever.

Not surprisingly, every proposal to do this has been found seriously flawed, and has led the American public to think of MX as an expensive fixation.

The bureaucracy's approach to strategic weapons has proved to be unsound not only strategically, but also politically. It has gone a long way toward destroying the American people's faith in their government's ability to defend them. The rest of us, however, are not obliged to follow this foolish path.

Rather, if we look at strategic weapons not as doomsday machines but realistically, even the MX has some limited, but very real, usefulness. All of our other strategic weapons were designed for more or less soft targets —

to kill Russians. But the MX, with the 475-kiloton warhead and accuracy better than Minuteman's one-tenth of a mile, would have an excellent chance of destroying not more Russian people—which should not be our objective in any event—but the Soviet missiles which threaten us.

As for the much-maligned Dense Pack, its untouted virtue is that, by forcing attacking warheads to pass through a narrow point in space and time, it makes possible an effective, relatively simple ballistic missile defense. An antimissile missile need only hit one spot in space to defend a whole missile field.

We must finally recognize that the flaws of Dense Pack, and of every other deployment scheme for land-based missiles, exist because increasing Soviet missile accuracy has narrowed our strategic options. Since nothing fixed and undefended can resist modern attacks, land-based missiles either must be launched on warning or must be made mobile. Launch on warning is too dangerous, too prone to error. So we must go mobile.

It was downright silly to put so many counterforce warheads on a single missile, and thereby to make the MX too big to move and hide easily. So, while we build a few MXs, we should be working quickly to build many small, mobile, hideable, highly accurate single-warhead ICBMs. By 1988 we could have a fleet of invulnerable missiles which, if we ever had to use them, would drastically reduce the Soviets' ability to do us harm.

Let us realize, however, that, useful as these counterforce missiles would be, they would not change the dread fact that, as a result of their enormous build-up during the past 15 years, the Soviets can still threaten to do more harm to our forces than we do to theirs. Also, because the Soviets have so many operating production lines for long-range missiles, the chances are great we will never be able to correct this imbalance by out-building them.

We should also realize that there is evidence that the Soviets are moving toward basing an ever-larger proportion of their ICBM force on mobile launchers. Our counterforce missiles may not be able to find them. As a practical matter, in order to minimize the number of Soviet warheads reaching the U.S., we are going to have to build systems to destroy Soviet missiles in flight.

I believe a variety of good defenses against ballistic missiles is possible. We should no more seek "the perfect" defensive system than we should seek "the perfect" basing mode for land-based missiles. None exists. When we insist on perfect solutions we wind up empty-handed. Rather we should build our defenses in layers, realizing a series of partial successes can add up to very substantial protection.

No defensive system, by itself, can do the whole job. If an attacker chooses to waste his weapons overwhelming one part of the system, he must lighten the attack on other parts. Moreover, prudent people depend on layers of defenses, Indeed, all ground-based ABMs would function much better if incoming attacks were thinned prior to coming into their range.

The technology of space-based lasers gives us substantial hope that attacking Soviet missiles could be defeated, or at the very least severely thinned, just after they rose out of the atmosphere.

Today, I am convinced, no informed person can any longer deny that the requisite laser power, the requisite accuracy in pointing and tracking, and the know-how for building the big space-based mirrors is in hand.

In 1980 a study for the Pentagon—led by a man vehemently opposed to building lasers—said that the U.S. could have a fleet of missile-killing lasers in space by 1994. The General Accounting Office, in a report released in 1982, said that the technology of space lasers was basically underfunded and that, with an intelligent increase of support, the U.S. could soon be testing a space-laser weapon.

Some have expressed fear that outer space might become militarized. This fear is misplaced. Space has been militarized for offensive purposes for a generation. We now have the choice of putting into space the means to destroy weapons of mass destruction. By building such means, we would tend to move the strategic struggle away from people, and into a realm where our technological superiority would be decisive. On what moral, political, or strategic grounds should we refuse to defend ourselves?

I'm afraid that the Department of Defense is failing to push for the several means of protecting Americans against ballistic missiles for the same reason that it preferred the MX over the small mobile counterforce missiles and for the same reason that it made the worst kind of arguments in favor of MX.

That reason is a commitment, more habitual than conscious, to the strategic utopianism of the 1960s: War can be banished forever if both the American and Soviet peoples are kept fully and forever vulnerable.

No one should be surprised, however, that both the Soviets and the American people have found this vision repelling. The Soviets, of course, are making frighteningly realistic preparations for both offense and defense. The time has arrived technically (if not intellectually) when we can focus our national effort not on greater destruction of mankind, but on the ever-greater protection of Americans from that greater destructive power.

# Space-Based Ballistic-Missile Defense

WITH RICHARD L. GARWIN, KURT GOTTFRIED
AND HENRY W. KENDALL

For two decades, both the U.S. and the U.S.S.R. have been vulnerable to a devastating nuclear attack, inflicted by one side on the other in the form of either a first strike or a retaliatory second strike. This situation did not come about as the result of careful military planning. "Mutual Assured Destruction" is not a policy or a doctrine but rather a fact of life. It simply descended like a medieval plague—a seemingly inevitable consequence of the enormous destructive power of nuclear weapons, of rockets that could hurl them across almost half of the globe in 30 minutes and of the impotence of political institutions in the face of such momentous technological innovations.

This grim development holds different lessons for different people. Virtually everyone agrees that the world must eventually escape from the shadow of mutual assured destruction, since few are confident that deterrence by threat of retaliation can avert a holocaust indefinitely. Beyond this point, however, the consensus dissolves. Powerful groups in the governments of both superpowers apparently believe that unremitting competition, albeit short of war, is the only realistic future one can plan for. In the face of much evidence to the contrary, they act as if the aggressive exploitation for military purposes of anything technology has to offer is critical to the security of the nation they serve. Others seek partial measures that could at least curb the arms race, arguing that this approach has usually been sidetracked by short-term (and shortsighted) military and political goals. Still others have placed varying degrees of faith in radical solutions: novel political moves, revolutionary technological advances, or some combination of the two.

President Reagan's Strategic Defense Initiative belongs in this last category. In his televised speech in 1983 calling on the nation's scientific community "to give us the means of rendering these nuclear weapons impotent and obsolete" the president expressed the hope that a technological revolution would enable the U.S. to "intercept and destroy strategic ballistic missiles before they reached our own soil or that of our allies." If such a breakthrough could be achieved, he said, "free people could live secure in the knowledge that their security did not rest upon the threat of instant U.S. retaliation."

Can this vision of the future ever become reality? Can any system for ballistic-missile defense eliminate the threat of nuclear annihilation? Would the quest for such a defense put an end to the strategic-arms race, as the president and his supporters have suggested, or is it more likely to accelerate that race? Does the president's program hold the promise of a secure and peaceful world or is it perhaps the most grandiose manifestation of the illusion that science can recreate the world that disappeared when the first nuclear bomb was exploded in 1945?

These are complex questions, with intertwined technical and political strands. They must be examined carefully before the U.S. commits itself to the quest for such a defense, because if the president's dream is to be pursued, space will become a potential field of confrontation and battle. It is partly for this reason the Strategic Defense Initiative is commonly known as the "Star Wars" program.

Our discussion of the political implications of the president's Strategic Defense Initiative will draw on the work of two of our colleagues, Peter A. Clausen of the Union for Concerned Scientists and Richard Ned Lebow of Cornell University.

The search for a defense against nuclear-armed ballistic missiles began three decades ago. In the 1960s both superpowers developed anti-ballistic-missile (ABM) systems based on the use of interceptor missiles armed with nuclear warheads. In 1968 the U.S.S.R. began to operate an ABM system around Moscow based on the Galosh interceptor, and in 1974 the U.S. completed a similar system to protect Minuteman missiles near Grand Forks Air Force Base in North Dakota. (The U.S. system was dismantled in 1975.)

Although these early efforts did not provide an effective defense against a major nuclear attack, they did stimulate two developments that have been dominant features of the strategic landscape ever since: the ABM Treaty of

1972 and the subsequent deployment of multiple independently targetable reentry vehicles (MIRVs), first by the U.S. and later by the U.S.S.R.

In the late 1960s a number of scientists who had been involved in investigating the possibility of ballistic-missile defense in their capacity as high-level advisers to the U.S. government took the unusual step of airing their criticism of the proposed ABM systems both in congressional testimony and in the press.* Many scientists participated in the ensuing debate, and eventually a consensus emerged in the scientific community regarding the flaws in the proposed systems.

The scientists' case rested on a technical assessment and a strategic prognosis. On the technical side they pointed out that the systems then under consideration were inherently vulnerable to deception by various countermeasures and to preemptive attack on their exposed components, particularly their radars. On the strategic side the scientists argued that the U.S.S.R. could add enough missiles to its attacking force to ensure penetration of any such defense. These arguments eventually carried the day, and they are still germane. They were the basis for the ABM Treaty, which was signed by President Nixon and General Secretary Brezhnev in Moscow in May 1972. The ABM Treaty formally recognized that not only the deployment but also the development of such defensive systems would have to be strictly controlled if the race in offensive missiles was to be contained.

MIRVs were originally conceived as the ideal countermeasure to ballistic-missile defense, and in a logical world they would have been abandoned with the signing of the ABM Treaty. Nevertheless, the U.S. did not try to negotiate a ban on MIRVs. Instead it led the way to their deployment in spite of repeated warnings by scientific advisers and the Arms Control and Disarmament Agency to senior government officials that MIRVs would undermine the strategic balance and ultimately be to the advantage of the U.S.S.R. because of its larger ICBMs. The massive increase in the number of nuclear warheads in both strategic arsenals during the 1970s is largely attributable to the introduction of MIRVs. The result, almost everyone now agrees, is a more precarious strategic balance.

The president's Strategic Defense Initiative is much more ambitious than the ABM proposals of the 1960s. To protect an entire society, a nationwide defense of "soft" targets such as cities would be necessary; in

---

*See "Antiballistic-Missile Systems" in this volume.

contrast, the last previous U.S. ABM plan—the Safeguard system proposed by the Nixon administration in 1969—was intended to provide only a "point" defense of "hard" targets such as missile silos and command bunkers. The latter mission could be accomplished by a quite permeable terminal-defense system that intercepted warheads very close to their targets, since a formidable retaliatory capability would remain even if most of the missile silos were destroyed. A large metropolitan area, on the other hand, could be devastated by a handful of weapons detonated at high altitude; if necessary, the warheads could be designed to explode on interception.

To be useful, a nationwide defense would have to intercept and eliminate virtually all the 10,000 or so nuclear warheads that each side is currently capable of committing to a major strategic attack. For a city attack, it could not wait until the atmosphere allowed the defense to discriminate between warheads and decoys. Such a high rate of attrition would be conceivable only if there were several layers of defense, each of which could reliably intercept a large percentage of the attacking force. In particular, the first defensive layer would have to destroy most of the attacking warheads soon after they left their silos or submerged submarines, while the booster rockets were still firing. Accordingly, boost-phase interception would be an indispensable part of any defense of the nation as a whole.

Booster rockets rising through the atmosphere thousands of miles from U.S. territory could be attacked only from space. That is why the Strategic Defense Initiative is regarded primarily as a space-weapons program. If the president's plan is actually pursued, it will mark a turning point in the arms race perhaps as significant as the introduction of ICBMs.

Several quite different outcomes of the introduction of space weapons have been envisioned. One view (apparently widely held in the Reagan administration) has been expressed most succinctly by Robert S. Cooper, director of the Defense Advanced Research Projects Agency. Testifying in 1983 before the Armed Services Committee of the House of Representatives, Cooper declared: "The policy for the first time recognizes the need to control space as a military environment." Indeed, given the intrinsic vulnerability of space-based systems, the domination of space by the U.S. would be a prerequisite to a reliable ballistic-missile defense of the entire nation. For that reason, among others, the current policy also calls for the acquisition by the U.S. of antisatellite weapons (see "Antisatellite Weapons," by Richard L. Garwin, Kurt Gottfried, and Donald L. Hafner, *Scientific American,* June 1984).

The notion that the U.S. could establish and maintain supremacy in space ignores a key lesson of the post-Hiroshima era: a technological breakthrough of even the most dramatic and unexpected nature can provide only a temporary advantage. Indeed, the only outcome one can reasonably expect is that both superpowers would eventually develop space-based ballistic-missile-defense systems. The effectiveness of these systems would be uncertain and would make the strategic balance more precarious than it is today. Both sides will have expanded their offensive forces to guarantee full confidence in their ability to penetrate defenses of unknown reliability, and the incentive to cut one's own losses by striking first in a crisis will be even greater than it is now. Whether or not weapons deployed in space could ever provide a reliable defense against ballistic missiles, they would be potent antisatellite weapons. As such they could be used to promptly destroy an opponent's early-warning and communications satellites, thereby creating a need for critical decisions at a tempo ill-suited to the speed of human judgment.

Our analysis of the prospects for a space-based defensive system against ballistic-missile attack will focus on the problem of boost-phase interception. It is not only an indispensable part of the currently proposed systems but also what distinguishes the current concept from all previous ABM plans. On the basis of our technical analysis and our assessment of the most likely response of the U.S.S.R., we conclude that the pursuit of the president's program would inevitably stimulate a large increase in the Russian strategic offensive forces, further reduce the chances of controlling events in a crisis and possibly provoke the nuclear attack it was designed to prevent. In addition, the reliability of the proposed defense would remain a mystery until the fateful moment at which it was attacked.

Before assessing the task of any defense, one must first examine the likely nature of the attack. In this case we shall concentrate on the technical and military attributes of the land-based ICBM and on how a large number of such missiles could be used in combination to mount a major strategic attack.

The flight of an ICBM begins when the silo door opens and hot gases eject the missile. The first-stage booster then ignites. After exhausting its fuel, the first stage falls away as the second stage takes over; this sequence is usually repeated at least one more time. The journey from the launch point to where the main rockets stop burning is the boost phase. For the present generation of ICBMs the boost phase lasts for three to five minutes and ends at an altitude of 300 to 400 kilometers, above the atmosphere.

A typical ICBM in the strategic arsenal of the U.S. or the U.S.S.R. is equipped with MIRVs, which are dispensed by a maneuverable carrier vehicle called a bus after the boost phase ends. The bus releases the MIRVs one at a time along slightly different trajectories toward their separate targets. If there were defenses, the bus could also release a variety of penetration aids, such as light-weight decoys, reentry vehicles camouflaged to resemble decoys, radar-reflecting wires called chaff and infrared-emitting aerosols. Once the bus had completed its task, the missile would be in midcourse. At that point the ICBM would have proliferated into a swarm of objects, each of which, no matter how light, would move along a ballistic trajectory indistinguishable from those of its accompanying objects. Only after the swarm reentered the atmosphere would the heavy, specially shaped reentry vehicles be exposed as friction with the air tore away the screen of lightweight decoys and chaff.

This brief account reveals why boost-phase interception would be crucial: every missile that survived boost phase would become a complex "threat cloud" by the time it reached midcourse. Other factors also amplify the importance of boost-phase interception. For one thing, the booster rocket is a much larger and more fragile target than the individual reentry vehicles are. For another, its flame is an abundant source of infrared radiation, enabling the defense to get an accurate fix on the missile. It is only during boost phase that a missile reveals itself by emitting an intense signal that can be detected at a large distance. In midcourse it must first be found by illuminating it with microwaves (or possibly laser light) and then sensing the reflected radiation, or by observing its weak infrared signal, which is due mostly to reflection of the earth's infrared radiation.

Because a nationwide defense must be capable of withstanding any kind of strategic attack, the exact nature of the existing offensive forces is immaterial to the evaluation of the defense. At present a full-scale attack by the U.S.S.R. on the U.S. could involve as many as 1,400 land-based ICBMs. The attack might well begin with submarine-launched ballistic missiles (SLBMs), since their unpredictable launch points and short flight times (10 minutes or less) would lend the attack an element of surprise that would be critical if the national leadership and the ground-based bomber force were high-priority targets.

SLBMs would be harder to intercept than ICBMs, which spend 30 minutes or so on trajectories whose launch points are precisely known. Moreover, a space-based defense system would be unable to intercept ground-hugging cruise missiles, which can deliver nuclear warheads to distant targets with an accuracy that is independent of range. Both superpowers are developing sea-launched cruise missiles, and these weapons are certain to become a major part of their strategic forces once space-based ballistic-missile-defense systems appear on the horizon.

The boost-phase layer of the defense would require many components that are not weapons in themselves. They would provide early warning of an attack by sensing the boosters' exhaust plumes; ascertain the precise number of the attacking missiles and, if possible, their identities; determine the trajectories of the missiles and get a fix on them; assign, aim and fire the defensive weapons; assess whether or not interception was successful, and, if time allowed, fire additional rounds. This intricate sequence of operations would have to be automated, because the total duration of the boost phase, now a few minutes, is likely to be less than 100 seconds by the time the proposed defensive systems are ready for deployment.

If a sizable fraction of the missiles were to survive boost-phase interception, the midcourse defensive layer would have to deal with a threat cloud consisting of hundreds of thousands of objects. For example, each bus could dispense as many as 100 empty aluminized Mylar balloons weighing only 100 grams each. The bus would dispense reentry vehicles (and possibly some decoy reentry vehicles of moderate weight) enclosed in identical balloons. The balloons and the decoys would have the same optical and microwave "signature" as the camouflaged warheads, and therefore the defensive system's sensors would not be able to distinguish between them. The defense would have to disturb the threat cloud in some way in order to find the heavy reentry vehicles, perhaps by detonating a nuclear explosive in the path of the cloud. To counteract such a measure, however, the reentry vehicles could be designed to release more balloons. Alternatively, the midcourse defense could be designed to target everything in the threat cloud, a prodigious task that might be beyond the supercomputers expected a decade from now. In short, the midcourse defense would be overwhelmed unless the attacking force was drastically thinned out in the boost phase.

Because the boosters would have to be attacked while they could not yet be seen from any point on the earth's surface accessible to the defense, the defensive system would have to initiate boost-phase interception from a point in space, at a range measured in thousands of kilometers. Two types of "directed energy" weapons are currently under investigation for this purpose: one type based on the use of laser beams, which travel at the speed of light (300,000 kilometers per second), and the other based on the use of particle beams, which are almost as fast. Nonexplosive projectiles that home on the booster's infrared signal have also been proposed.

There are two alternatives for basing such weapons in space. They could be in orbit all the time or they could be "popped up" at the time of the attack. There are complementary advantages and disadvantages to each approach. With enough weapons in orbit, some would be "on station" whenever they were needed, and they could provide global coverage; on the other hand, they would be inefficient because of the number of weapons that would have to be actively deployed, and they would be extremely vulnerable. Pop-up weapons would be more efficient and less vulnerable, but they would suffer from formidable time constraints and would offer poor protection against a widely dispersed fleet of strategic submarines.

Pop-up interceptors of ICBMs would have to be launched from submarines, since the only accessible points close enough to the Russian ICBM silos are in the Arabian Sea and the Norwegian Sea, at a distance of more than 4,000 kilometers. An interceptor of this type would have to travel at least 940 kilometers before it could "see" an ICBM just burning out at an altitude of 200 kilometers. If the interceptor were lofted by an ideal instant-burn booster with a total weight-to-payload ratio of 14 to 1, it could reach the target-sighting point in about 120 seconds. For comparison, the boost phase of the new U.S. MX missile (which has a weight-to-payload ratio of 25 to 1) is between 150 and 180 seconds. In principle, therefore, it should just barely be possible by this method to intercept a Russian missile comparable to the MX, provided the interception technique employed a beam that moves at the speed of light. On the other hand, it would be impossible to intercept a large number of missiles, since many silos would be more than 4,000 kilometers away, submarines cannot launch all their missiles simultaneously and 30 seconds would leave virtually no time for the complex sequence of operations the battle-management system would have to perform.

A report prepared for the Fletcher panel, the study team set up last year by the Department of Defense under the chairmanship of James C. Fletcher of the University of Pittsburgh to evaluate the Strategic Defense Initiative for the president, bears on this question. According to the report, it is possible to build ICBMs that could complete the boost phase and disperse their MIRVs in only 60 seconds, at a sacrifice of no more than 20 percent of payload. Even with zero-decision-time a hypothetical instant-burn rocket that could pop up an interceptor system in time for a speed-of-light attack on such an ICBM would need an impossible weight-to-payload ratio in excess of 800 to 1! Accordingly all pop-up interception schemes, no matter what kind of antimissile weapon they employ, depend on the assumption that the U.S.S.R. will not build ICBMs with a boost phase so short that no pop-up system could view the burning booster.

The time constraint faced by pop-up schemes could be avoided by putting at least some parts of the system into orbit. An antimissile satellite in a low orbit would have the advantage of having the weapon close to its targets, but it would suffer from the "absentee" handicap: because of its own orbital motion, combined with the earth's rotation, the ground track of such a satellite would pass close to a fixed point on the earth's surface only twice a day. Hence for every low-orbit weapon that was within range of the ICBM silos many others would be "absentees": they would be below the horizon and unable to take part in the defense. This unavoidable replication would depend on the range of the defensive weapon, the altitude and inclination of its orbit and the distribution of the enemy silos.

The absentee problem could be solved by mounting at least some components of the defensive system on a geosynchronous satellite, which remains at an altitude of some 36,000 kilometers above a fixed point on the Equator, or approximately 39,000 kilometers from the Russian ICBM fields. Whichever weapon were used, however, this enormous range would make it virtually impossible to exploit the radiation from the booster's flame to accurately fix an aim point on the target. The resolution of any optical instrument, whether it is an observing telescope or a beam-focusing mirror, is limited by the phenomenon of diffraction. The smallest spot on which a mirror can focus a beam has a diameter that depends on the wavelength of the radiation, the aperture of the instrument and the distance to the spot. For infrared radiation from the booster's flame the wavelength would typically be one micrometer, so that targeting on a spot 50 centimeters across at a range of 39,000 kilometers would require a precisely shaped mirror 100 meters across—roughly the length of a football field. (For comparison, the largest telescope mirrors in the world today are on the order of five meters in diameter.)

The feasibility of orbiting a high-quality optical instrument of this stupendous size seems remote. The wavelengths used must be shortened, or the viewing must be reduced, or both. Accordingly it has been suggested that a geosynchronous defensive system might be augmented by other optical elements deployed in low orbits.

One such scheme that has been proposed calls for an array of ground-based excimer lasers designed to work in conjunction with orbiting optical elements. The excimer laser incorporates a pulsed electron beam to excite a mixture of gases such as xenon and chlorine into a metastable molecular state, which spontaneously reverts to the molecular ground state; the latter in turn immediately dissociates into two atoms, emitting the excess energy in the form of ultraviolet radiation at a wavelength of .3 micrometer.

Each ground-based excimer laser would send its beam to a geosynchronous mirror with a diameter of five meters, and the geosynchronous mirror would in turn reflect the beam toward an appropriate "fighting mirror" in low orbit. The fighting mirror would then redirect and concentrate the beam onto the rising booster rockets, depending on an accompanying infrared telescope to get an accurate fix on the boosters.

The main advantage of this scheme is that the intricate and heavy lasers, together with their substantial power supplies, would be on the ground rather than in orbit. The beam on any ground-based laser, however, would be greatly disturbed in an unpredictable way by ever present fluctuations in the density of the atmosphere, causing the beam to diverge and lose its effectiveness as a weapon.

Assuming that such a system could be made to work perfectly, its power requirement can be estimated. Such an exercise is illuminating because it gives an impression of the staggering total cost of the system. Again information from the Fletcher panel provides the basis for our estimate. Apparently the "skin" of a booster can be "hardened" to withstand an energy deposition of 200 megajoules per square meter, which is roughly what is required to evaporate a layer of carbon three millimeters thick. With the aid of a geosynchronous mirror five meters in diameter and a fighting and viewing mirror of the same size, the beam of the excimer laser described above would easily be able to make a spot one meter across on the skin of a booster at a range of 3,000 kilometers from the fighting mirror; the resulting lethal dose would be about 160 megajoules.

A successful defense against an attack by the 1,400 ICBMs in the current Russian force would require a total energy deposition of 225,000 megajoules. (A factor of about 10 is necessary to compensate for atmospheric absorption, reflection losses at the mirrors and overcast skies.) If the time available for interception were 100 seconds and the lasers had an electrical efficiency of 6 percent, the power requirement would be more than the output of three-hundred 1,000-megawatt power plants, or more than 60 percent of the current electrical generating capacity of the entire U.S. Moreover, this energy could not be extracted instantaneously from the national power grid, and it could not be stored by any known technology for instantaneous discharge. Special power plants would have to be built; even though they would need to operate only for minutes an investment of $300 per kilowatt is a reasonable estimate, and so the outlay for the power supply alone would exceed $100 billion.

This partial cost estimate is highly optimistic. It assumes that all the boosters could be destroyed on the first shot, that the Russians would not have shortened the boost phase on their ICBMs, enlarged their total strategic-missile force or installed enough countermeasures to degrade the defense significantly by the time this particular defensive system was ready for deployment at the end of the century. Of course the cost of the entire system of lasers, mirrors, sensors and computers would far exceed the cost of the power plant, but at this stage virtually all the required technologies are too immature to allow a fair estimate of their cost.

The exact number of mirrors in the excimer scheme depends on the intensity of the laser beams. For example, if the laser could deliver a lethal dose of heat in just five seconds, one low-orbit fighting mirror could destroy 20 boosters in the assumed time of 100 seconds. It follows that 70 mirrors would have to be within range of the Russian silos to handle the entire attack, and each mirror would need to have a corresponding mirror in a geosynchronous orbit. If the distance at which a fighting mirror could focus a small enough spot of light was on the order of 3,000 kilometers, there would have to be about six mirrors in orbit elsewhere for every one "on station" at the time of the attack, for a total of about 400 fighting mirrors. This allowance for absenteeism is also optimistic, in that it assumes the time needed for targeting would be negligible, there would be no misses, the Russian countermeasures would be ineffective and excimer lasers far beyond the present state-of-the-art could be built.

The second boost-phase interception scheme we shall consider is a pop-up system based on the X-ray laser, the only known device light enough to be a candidate for this role. As explained above, shortening the boost phase of the attacking missiles would negate any pop-up scheme. In this case a shortened boost phase would be doubly crippling, since the booster would stop burning within the atmosphere, where X rays cannot penetrate. Nevertheless, the X-ray laser has generated a good deal of interest, and we shall consider it here even though it would be feasible only if the Russians were to refrain from adapting their ICBMs to thwart this threat.

The X-ray laser consists of a cylindrical array of thin fibers surrounding a nuclear explosive. The thermal X rays generated by the nuclear explosion stimulate the emission of X-radiation from the atoms in the fibers. The light produced by an ordinary optical laser can be highly collimated, or directed, because it is reflected back and forth many times between the mirrors at the ends of the laser. An intense X-ray beam, however, cannot be reflected in this way, and so the proposed X-ray laser would emit a rather divergent beam; for example, at a distance of 4,000 kilometers it would make a spot about 200 meters across.

The U.S. research program on X-ray lasers is highly classified. According to a Russian technical publication, however, such a device can be expected to operate at an energy of about 1,000 electron volts. Such a "soft" X-ray pulse would be absorbed in the outermost fraction of a micrometer of a booster's skin, "blowing off" a thin surface layer. This would have two effects. First, the booster as a whole would recoil. The inertial-guidance system would presumably sense the blow, however, and it could still direct the warheads to their targets. Second, the skin would be subjected to an abrupt pressure wave that, in a careless design, could cause the skin to shear at its supports and damage the booster's interior. A crushable layer installed under the skin could prolong and weaken the pressure wave, however, thereby protecting both the skin and its contents.

Other interception schemes proposed for ballistic-missile defense include chemical-laser weapons, neutral-particle-beam weapons and nonexplosive homing vehicles, all of which would have to be stationed in low orbits.

The brightest laser beam attained so far is an infrared beam produced by a chemical laser that utilizes hydrogen fluoride. The U.S. Department of Defense plans to demonstrate a two-megawatt version of this laser by 1987. Assuming that 25-megawatt hydrogen-fluoride lasers and optically perfect 10-meter mirrors eventually become available, a weapon with a "kill radius" of 3,000 kilometers would be at hand. A total of 300 such lasers in low orbits could destroy 1,400 ICBM boosters in the absence of countermeasures if every component worked to its theoretical limit.

A particle-beam weapon could fire a stream of energetic charged particles, such as protons, that could penetrate deep into a missile and disrupt the semiconductors in its guidance system. A charged-particle beam, however, would be bent by the earth's magnetic field and therefore could not be aimed accurately at distant targets. Hence any plausible particle-beam weapon would have to produce a neutral beam, perhaps one consisting of hydrogen atoms (protons paired with oppositely charged electrons). This could be done, although aiming the beam would still present formidable problems. Interception would be possible only above the atmosphere at an altitude of 150 kilometers or more, since collisions with air molecules would disintegrate the atoms and the geomagnetic field would then fan out the beam. Furthermore, by using gallium arsenide semiconductors, which are about 1,000 times more resistant to radiation damage than silicon semiconductors, it would be possible to protect the missile's guidance computer from such a weapon.

Projectiles that home on the booster's flame are also under discussion. They have the advantage that impact would virtually guarantee destruction, whereas a beam weapon would have to dwell on the fast-moving booster for some time. Homing weapons, however, have two drawbacks that preclude their use as boost-phase interceptors. First, they move at less than .01 percent of the speed of light, and therefore they would have to be deployed in uneconomically large numbers. Second, a booster that burned out within the atmosphere would be immune to them, since friction with the air would blind their homing sensors.

That such a homing vehicle can indeed destroy an object in space was demonstrated by the U.S. Army in its current Homing Overlay test series. On June 10 a projectile launched from Kwajalein Atoll in the Pacific intercepted a dummy Minuteman warhead at an altitude of more than 100 miles. The interceptor relied on a homing technique similar to that of the Air Force's aircraft-launched antisatellite weapon. The debris from the collision was scattered over many tens of kilometers and was photographed by tracking telescopes. The photographs show, among other things, the difficulty of evading a treaty that banned tests of weapons in space.

In an actual ballistic-missile-defense system, such an interceptor might have a role in midcourse defense. It would have to be guided to a disguised reentry vehicle hidden in a swarm of decoys and other objects designed to confuse its infrared sensors. The potential of this technique for midcourse interception remains to be demonstrated, whereas its potential for boost-phase interception is questionable in view of the considerations mentioned above. On the other hand, a satellite is a larger and more fragile target than a reentry vehicle, and so the recent test shows the U.S. has a low-altitude antisatellite capability at least equivalent to the U.S.S.R.'s.

The importance of countermeasures in any consideration of ballistic-missile defense was emphasized recently by Richard D. DeLauer, Undersecretary of Defense for Research and Engineering. Testifying on this subject before the House Armed Services Committee, DeLauer stated that "any defensive system can be overcome with proliferation and decoys, decoys, decoys, decoys."

One extremely potent countermeasure has already been mentioned, namely that shortening the boost phase of the offensive missiles would nullify any boost-phase interception scheme based on X-ray lasers, neutral-particle beams or homing vehicles. Many other potent countermeasures that exploit existing technologies can also be envisioned. All of them rely on generic weaknesses of the defense. Among these weaknesses four stand out: (1) Unless the defensive weapons were cheaper than the offensive ones, any defense could simply be overwhelmed by a missile buildup; (2) the defense would have to attack every object that behaves like a booster; (3) any space-based defensive component would be far more vulnerable than the ICBMs it was designed to destroy; (4) since the booster, not the flame, would be the target, schemes based on infrared detection could be easily deceived.

Countermeasures can be divided into three categories: those that are threatening, in the sense of manifestly increasing the risk to the nation deploying the defensive system; those that are active, in the sense of attacking the defensive system itself; and those that are passive, in the sense of frustrating the system's weapons. These distinctions are politically and psychologically significant.

The most threatening response to a ballistic-missile-defense system is also the cheapest and surest: a massive buildup of real and fake ICBMs. The deployment of such a defensive system would violate the ABM Treaty, almost certainly resulting in the removal of all negotiated constraints on offensive missiles. Therefore many new missile silos could be constructed. Most of them could be comparatively inexpensive fakes arrayed in clusters about 1,000 kilometers across to exacerbate the satellites' absentee problem. The fake silos could house decoy ICBMs—boosters without expensive warheads or guidance packages—that would be indistinguishable from real ICBMs during boost phase. An attack could begin with a large proportion of decoys and shift to real ICBMs as the defense exhausted its weapons.

All space systems would be highly vulnerable to active countermeasures. Few targets could be more fragile than a large, exquisitely made mirror whose performance would be ruined by the slightest disturbance. If an adversary were to put a satellite into the same orbit as that of the antimissile weapon but moving in the opposite direction, the relative velocity of the two objects would be about 16 kilometers per second, which is eight times faster than that of a modern armor-piercing antitank projectile. If the satellite were to release a swarm of one-ounce pellets, each pellet could penetrate 15 centimeters of steel (and much farther if it were suitably shaped). Neither side could afford to launch antimissile satellites strong enough to withstand such projectiles. Furthermore, a large number of defensive satellites in low or geosynchronous orbits could be attacked simultaneously by "space mines": satellites parked in orbit near their potential victims and set to explode by remote control or when tampered with.

Passive countermeasures could be used to hinder targeting or to protect the booster. The actual target would be several meters above the flame, and the defensive weapon would have to determine the correct aim point by means of an algorithm stored in its computer. The aim point could not be allowed to drift by more than a fraction of a meter, because the beam weapon would have to dwell on one spot for at least several seconds as the booster moved several tens of kilometers. Aiming could therefore be impeded if the booster flame were made to fluctuate in an unpredictable way. This effect could be achieved by causing additives in the propellant to be emitted at random from different nozzles or by surrounding the booster with a hollow cylindrical "skirt" that could hide various fractions of the flame or even move up and down during boost phase.

Booster protection could take different forms. A highly reflective coating kept clean during boost phase by a strippable foil wrapping would greatly reduce the damaging effect of an incident laser beam. A hydraulic cooling system or a movable heat-absorbing ring could protect the attacked region at the command of heat sensors. Aside from shortening the boost phase the attacking nation could also equip each booster with a thin metallic sheet that could be unfurled at a high altitude to absorb and deflect an X-ray pulse.

Finally, as DeLauer has emphasized, all the proposed space weapons face formidable systemic problems. Realistic testing of the system as a whole is obviously impossible and would have to depend largely on computer simulation. According to DeLauer, the battle-management system would face a task of prodigious complexity that is "expected to stress software-development technology"; in addition it would have to "operate reliably even in the presence of disturbances caused by nuclear-weapons effects or direct-energy attack." The Fletcher panel's report states that the "*survivability of the system components is a critical issue whose resolution requires a combination of technologies and tactics that remain to be worked out.*" Moreover, nuclear attacks need not be confined to the battle-management system. For example, airbursts from a precursor salvo of SLBMs could produce atmospheric disturbances that would cripple an entire defensive system that relied on the ground-based laser scheme.

Spokesmen for the Reagan administration have stated that the Strategic Defense Initiative will produce a shift to a "defense-dominated" world. Unless the move toward ballistic-missile defense is coupled with deep cuts in both sides' offensive forces, however, there will be no such shift. Such a coupling would require one or both of the following conditions: a defensive technology that was so robust and cheap that countermeasures or an offensive buildup would be futile, or a political climate that would engender arms-control agreements of unprecedented scope. Unfortunately neither of these conditions is in sight.

What shape, then, is the future likely to take if attempts are made by the U.S. and the U.S.S.R. to implement a space-based system aimed at thwarting a nuclear attack? Several factors will have a significant impact. First, the new technologies will at best take many years to develop, and, as we have argued, they will remain vulnerable to known countermeasures. Second, both sides are currently engaged in "strategic modernization" programs that will further enhance their already awesome offensive forces. Third, in pursuing ballistic-missile defense both sides will greatly increase their currently modest antisatellite capabilities. Fourth, the ABM Treaty, which is already under attack, will fall by the wayside.

These factors, acting in concert, will accelerate the strategic-arms race and simultaneously diminish the stability of the deterrent balance in a crisis. Both superpowers have always been inordinately sensitive to real and perceived shifts in the strategic balance. A defense that could not fend off a full-scale strategic attack but might be quite effective against a weak retaliatory blow following an all-out preemptive strike would be particularly provocative. Indeed, the leaders of the U.S.S.R. have often stated that any U.S. move toward a comprehensive ballistic-missile-defense system would be viewed as an attempt to gain strategic superiority, and that no effort would be spared to prevent such an outcome. It would be foolhardy to ignore these statements.

The most likely Russian response to a U.S. decision to pursue the president's Strategic Defense Initiative should be expected to rely on traditional military "worst case" analysis; in this mode of reasoning one assigns a higher value to the other side's capabilities than an unbiased examination of the evidence would indicate, while correspondingly undervaluing one's own capabilities. In this instance the Russians will surely overestimate the effectiveness of the U.S. ballistic-missile defense and arm accordingly. Many near-term options would then be open to them. They could equip their large SS-18 ICBMs with decoys and many more warheads; they could retrofit their deployed ICBMs with protective countermeasures; they could introduce fast-burn boosters; they could deploy more of their current-model ICBMs and sea-launched cruise missiles. The latter developments would be perceived as unwarranted threats by U.S. military planners, who would be quite aware of the fragility of the nascent U.S. defensive system. A compensating U.S. buildup in offensive missiles would then be inevitable. Indeed, even if both sides bought identical defensive systems from a third party, conservative military analysis would guarantee an accelerated offensive-arms race.

Once one side began to deploy space-based antimissile beam weapons, the level of risk would rise sharply. Even if the other side did not overrate the system's antimissile capability, it could properly view such a system as an immediate threat to its strategic satellites. A strategy of "launch on warning" might then seem unavoidable, and attempts might also be made to position space mines alongside the antimissile weapons. The last measure might in itself trigger a conflict since the antimissile system should be able to destroy a space mine at a considerable distance if it has any capability for its primary mission. In short, in a hostile political climate even a well-intentioned attempt to create a strategic defense could provoke war, just as the mobilizations of 1914 precipitated World War I.

Even if the space-based ballistic-missile defense did not have a cataclysmic birth, the successful deployment of such a defense would create a highly unstable strategic balance. It is difficult to imagine a system more likely to induce catastrophe than one that requires critical decisions by the second, is itself untested and fragile and yet is threatening to the other side's retaliatory capability.

In the face of mounting criticism, administration spokesmen have in recent months offered less ambitious rationales for the Strategic Defense Initiative than the president's original formulation. One theme is that the program is just a research effort and that no decision to deploy will be made for many years. Military research programs are not normally announced from the Oval Office, however, and there is no precedent for a $26-billion, five-year military-research program without any commitment to deployment. A program of this magnitude, launched under such auspices, is likely to be treated as an essential military policy by the U.S.S.R. no matter how it is described in public.

Another more modest rationale of the Strategic Defense Initiative is that it is intended to enhance nuclear deterrence. That role, however, would require only a terminal defense of hard targets, not weapons in space. Finally, it is contended that even an imperfect antimissile system would limit damage to the U.S.; the more likely consequence is exactly the opposite, since it would tend to focus the attack on cities, which could be destroyed even in the face of a highly proficient defense.

In a background report titled, *Directed Energy Missile Defense in Space*, released earlier in 1984 by the Congressional Office of Technology Assessment, the author, Ashton B. Carter of the Massachusetts Institute of Technology, a former Defense Department analyst with full access to classified data on such matters, concluded that "the prospect that emerging 'Star Wars' technologies, when further developed, will provide a perfect or near-perfect defense system . . . is so remote that it should not serve as the basis of public expectation or national policy." Based on our assessment of the technical issues, we are in complete agreement with this conclusion.

In our view, the questionable performance of the proposed defense, the ease with which it could be overwhelmed or circumvented and its potential as an antisatellite system would cause grievous damage to the security of the U.S. if the Strategic Defense Initiative were to be pursued. The path toward greater security lies in quite another direction. Although research, on ballistic-missile defense should continue at the traditional level of expenditure and within the constraints of the ABM Treaty, every effort should be made to negotiate a bilateral ban on the testing and use of space weapons.

It is essential that such an agreement cover all altitudes, because a ban on high-altitude antisatellite weapons alone would not be viable if directed-energy weapons were developed for ballistic-missile defense. Once such weapons were tested against dummy boosters or reentry vehicles at low altitude, they would already have the capability of attacking geosynchronous satellites without testing at high altitude. The maximum energy density of any such beam in a vacuum is inversely proportional to the square of the distance. Once it is demonstrated that such a weapon can deliver a certain energy dose in one second at a range of 4,000 kilometers, it is established that the beam can deliver the same dose at a range of 36,000 kilometers in approximately 100 seconds. Since the beam could dwell on a satellite indefinitely, such a device could be a potent weapon against satellites in geosynchronous orbits, even if it failed in its ballistic-missile-defense mode.

As mentioned above, the U.S. interception of a Minuteman warhead over the Pacific shows that both sides now have a ground-based antisatellite weapon of roughly equal capability. Hence there is no longer an asymmetry in such antisatellite weapons. Only a lack of political foresight and determination blocks the path to agreement. Such a pact would not permanently close the door on a defense-dominated future. If unforeseen technological developments were to take place in a receptive international political climate in which they could be exploited to provide greater security than the current condition of deterrence by threat of retaliation provides, the renegotiation of existing treaties could be readily achieved.

# The Technological Imperative

The industrial revolution, and hence our present material well-being, was founded on technology. In its early days and for two centuries thereafter it was important to develop everything that was technologically possible. Now there are so many possibilities that choices must be made. In the civilian economy, these choices are generally made by the marketplace.

In war, likewise, technological superiority has counted since ancient times. There are many examples in previous wars, but most impressive was World War II. The Allies' superiority in radar was decisive, both in averting disaster and in ultimate victory.

World War II ended with Hiroshima and Nagasaki. Albert Einstein summed up the situation after the atomic bombs had been dropped: "Everything has changed, except human thinking."

For a brief period, the U.S. government's thinking did change. President Harry Truman in 1946 appointed a committee chaired by David Lilienthal to explore the possibilities of international control of all activities related to atomic weapons and atomic power. When the positive report of this committee was endorsed by then Undersecretary of State Dean Acheson, President Truman proposed, through Bernard Baruch, that the United Nations establish an international control agency. The Soviet Union, having worked on military applications of nuclear fission since 1943, turned down the proposal.

After this, however, human thinking returned to pre-Hiroshima patterns. The atomic bomb was incorporated into our arsenal as if it were just another weapon. The Air Force built squadrons of bombers equipped with atomic bombs, so that soon there were hundreds of them. No thought was given to the demonstrable fact that just one atomic bomb was enough to devastate a city.

For several years the United States had a monopoly in atomic weapons. Then in 1949 the Soviets developed and tested their own. The test was detected by U.S. planes equipped to pick up any radioactive debris that might be in the air. Many scientists (and others) had expected that the Soviets would, sooner or later, develop an atomic weapon. For instance, in 1945 Fred Seitz and I published an article saying that "a determined country will be able to develop an atomic weapon in five years." As a matter of fact, it took the Soviets only four.

A number of American scientists felt in 1949 that it was necessary to stay ahead of the Soviets and, to this end, that the United States should develop the hydrogen bomb. They found willing ears in Congress and in some parts of the administration. President Truman was bombarded with arguments on both sides of the question, but the decisive news, which moved him to approve the development of the H-bomb, was the discovery of the treason of Klaus Fuchs. When it was shown that Fuchs had given the Soviets most of the information he had about various parts of the Manhattan Project— including whatever knowledge then existed about the possibility of hydrogen bombs—Truman decided that it had become a technological imperative for the United States to go ahead with the development of the vastly more powerful weapon.

It soon became apparent that this task was not nearly as easy as had been anticipated; the methods which had previously been considered simply were not promising. A new method had to be found, and it was Edward Teller who devised one in the spring of 1951. As is well known, that method was successfully tested in November 1952. The Soviets tested a preliminary device in August 1953, and a more developed one at the end of 1955. It is a matter of controversy whether they would have developed an H-bomb if we had not done so.

Was Truman's decision really a technological imperative? What if his decision had been negative and then the Soviets had confronted us with a test of their own H-bomb? A possible solution was suggested just before our test in 1952, and was taken up again in 1983 by McGeorge Bundy in the *New York Review of Books*. We could have announced that we would do the research leading to an H-bomb but would not test it. A hydrogen bomb (in contrast to an atomic bomb) is sufficiently complicated that nobody would consider adding it to the weapons stockpile without first testing it. And any test of an H-bomb in the atmosphere can be detected around the world.

An announcement of this kind would have led to two possibilities: either the world would have been spared this most devastating weapon, or some other country might have tested an H-bomb. Had such a test occurred, the United States would have quickly followed suit. We would then have been ready to develop the H-bomb as a weapon; and because of our great technological capacity, we probably would have obtained the ready-to-use weapon earlier than other countries. So it would have been possible for us to avoid this escalation in the arms race.

There would have been problems with this alternative path, one of them being the morale of the weapons laboratories. It is very discouraging to spend two or three years on the development of a completely new concept, and then to find that it could not even be tested, let alone accepted into the U.S. weapons arsenal. I know this disappointment well, having worked at both a weapons laboratory and an industrial laboratory concerned with development of missiles. When the missile designed by the industrial laboratory was rejected by the Air Force in favor of one developed by another company, it was a most discouraging blow. However, if we wish to escape the vicious cycle of ever-increasing armaments, we have to find a way to make weapons laboratories operate without the certainty that what they develop will actually be used.

The next important step in the armaments race was the intercontinental ballistic missile. The Soviets were the first to test such a missile in 1957. Their test of an ICBM was soon followed by Sputnik, the first artificial satellite. We made a frantic effort to catch up, and I think that in this case it was indeed justified to feel a technological imperative. But then there was the question of how many we should deploy. Then Secretary of Defense Robert McNamara decided on 1,000, in keeping with the existing number of bombing planes and atomic bombs. Soon we found out through our intelligence satellites that the Soviets had deployed only a very few ICBMs. Knowing this, it would have been sensible for us to reduce the number of ICBMs to around 200, which might have mitigated the arms race in missiles. But once we had installed 1,000, it was obvious that the Soviets would follow their technological imperative and build a similar number.

The first ICBMs carried a single warhead. This made for a rather stable balance: If either the Soviet Union or the United States were to attempt a first strike by attacking the other country's ICBM silos, the attacking nation could at best expect to destroy one of the other country's ICBMs for every one of its own. In fact, since surely not all of them would hit their targets, making a first strike would actually be a disadvantage in the case of two evenly matched adversaries.

This "happy" stability was disturbed by the Soviets' technological imperative which led them to develop antiballistic-missile (ABM) systems. "We can hit a fly in space," said Khrushchev, and an ABM system was deployed around Moscow. In response, the United States developed penetration aids (decoys, chaff, and so on) and increased the number of its ICBMs targeted on Moscow, thus negating the protection given by the ABM.

But U.S. designers were sure they could do better; they could put several warheads on one missile. In this way, the Soviets could not tell whether the swarm of objects coming at Moscow or some other target was one warhead and many decoys, or whether it contained perhaps several warheads, all of which had to be engaged by their ABMs. With this argument, MIRVing (multiple independently targeted reentry vehicles) was sold to U.S. decision-makers—and its development became a technological imperative.

Fairly soon thereafter, in 1972, the ABM Treaty was concluded. In it the Soviet Union and the United States agreed that neither of them would build more than two ABM systems and that each of these systems would be limited in the number of antiballistic missiles. Once that treaty was concluded, the United States would have been well advised to give up MIRVs. In fact, the Arms Control and Disarmament Agency warned that if we proceeded with building MIRVs the Soviets would follow suit, and they were much better equipped to do so because their missiles were much heavier than ours, and could therefore take a larger number of MIRVs. But no serious attempt was made to prohibit MIRV by means of a treaty with the Soviet Union.

Once we—and the Soviets—had MIRV, there was no longer any strategic stability in the ICBM system. A potential aggressor could, with just a few of his missiles, wipe out a large number of the opponent's silos. Thus the development and installation of MIRV was another example of the technological imperative being followed without regard for the consequences. MIRV reduced the security of both sides because a first-strike counterforce attack was now, in principle, possible. In fact, many U.S. strategic planners in the early 1980s feared that our missile silos had become vulnerable to a Soviet attack, and they spoke of a "window of vulnerability." However, the Scowcroft Commission, appointed by President Reagan in 1983, concluded that while our ICBMs might indeed become vulnerable one had to consider the whole triad of our strategic weapons: missile-carrying submarines remained invulnerable, and therefore a Soviet first strike against our silos would still not disarm us.

The technological imperative is again upon us. In the last years much progress had been made in such areas as heat-seeking missiles, electro-optics, lasers (including X-ray lasers), and computers. On the basis of this progress, many claims have been made by scientists and engineers that a defense against missiles is now possible. Persuaded by these claims, President Reagan, on March 23, 1983, launched his Strategic Defense Initiative (SDI), popularly known as "Star Wars." The intention of this project is to intercept ballistic missiles before they reach their targets, and thus gradually eliminate the threat of nuclear weapons carried by such missiles. Unfortunately, I and most of my colleagues who have looked into this problem are convinced that none of the proposed systems will work as advertised. They can, in fact, be defeated easily by countermeasures which will cost much less than it will cost to develop SDI. The technological imperative, however, is strong; our government feels that we must use this new technology in an attempt to reduce the threat of nuclear war. On the other hand, the Soviet Union has announced that it considers the project a threat to them, and therefore, unless negotiations can convince the Soviets to the contrary, SDI will lead to a new escalation in offensive weapons rather than to a decrease.

One of the pressures for developing the Strategic Defense Initiative comes from the weapons laboratories. Having recognized that there is little more to be done in improving offensive weapons, they are enthusiastic advocates of defensive weapons. And in this advocacy, they are finding a very receptive government.

An argument which is sometimes made is that the development of weapons technology is needed to stay ahead in technology as such, and that it will have a beneficial influence on peacetime technology. This surely has been the case in some instances, such as radar, which is now an essential navigation aid for commercial shipping and airlines. Likewise, the development of satellites had brought great civilian benefits. But today military technology demands such enormously sophisticated devices that it is highly unlikely that they will benefit the civilian economy to a substantial degree. On the contrary, the diversion of much scientific and engineering talent to military inventions has impoverished our civilian technology to a considerable extent. One reason for the great superiority of Japanese civilian technology in many fields is that the Japanese do not need to worry about military problems. I think that the redirection of engineering talents to civilian problems could very strongly benefit our own economy.

Whenever the technological imperative calls for new military devices, we should think very carefully about whether this particular development will contribute to the security of the United States and the world. In the case of MIRV it is clear that it did the opposite. I still consider the H-bomb a calamity. However, the nuclear submarine, whose "technological imperative" I did not recognize at the beginning of its development, has proved to be the best defense we have: ballistic-missile-launching submarines are the most invulnerable of our strategic weapons.

Preliminary considerations of the possible effects of new weapons should ideally take place at the level of the weapons laboratory. But the most important place for careful consideration of the possible effects of such systems is, of course, at the government level. A highly qualified group for such deliberations used to be the President's Science Advisory Committee. The *ad hoc* commissions appointed by President Reagan can also function well, provided their conclusions are not conditioned by previous decisions. Whatever group deliberates on the worth of a new weapons system should show restraint and should not lightly follow the technological imperative.

# Reducing the Risk of Nuclear War

## WITH ROBERT S. McNAMARA

Throughout history war has been the final arbiter of disputes and a finite disaster. Unbounded calamities—the apocalypse, Armageddon—were left for mythology. Forty years ago Hiroshima put an end to that distinction. This insight was expressed with exceptional clarity by President Reagan when he said that "a nuclear war cannot be won and must never be fought." And yet both superpowers' policies rely on thinking that is mired in the pre-nuclear past. Each strives ceaselessly to improve its arsenal and lays plans for fighting the war that must never be fought.

Although the risk of war between East and West seems low at present, should a military confrontation occur, the chance that it would escalate to all-out nuclear war is very, very high. That danger will haunt us as long as we persist on our present course. The combination of these factors—a high probability that war would destroy our society, and an indefinitely long exposure to that danger—produces a risk that is unacceptable. There is a widespread intuitive awareness of this peril.

When the president proposes a "Star Wars" space defense that would make nuclear weapons "impotent and obsolete," there is, therefore, an understandable outpouring of public support, even though most technical experts, inside the government and out, consider his proposal to be a nostalgic dream without a discernible connection to the realities of nuclear physics.

How, then, are we to escape our predicament? By heeding Einstein's admonition that "the unleashed power of the atom has changed everything save our modes of thinking." If we clearly face the implications of nuclear weapons, we will see the path through the hazards that science has forever unveiled. The path is not easy, but if it is followed with persistence, the risk of nuclear war will constantly recede and confidence that we are masters of our fate will be rebuilt.

The first large step toward our goal can be taken at the Geneva arms talks. It should be possible to pursue a ballistic-missile-defense research program, as desired by the president, and at the same time to negotiate a strengthening of the Antiballistic-Missile Treaty. By that means, the U.S. and Soviet positions on space defense could be reconciled and the way thereby opened to sharp cuts in offensive forces shaped in a manner that would lead to a much safer world in the twenty-first century.

In developing this thesis we will discuss:

• The situation today: a world with tens of thousands of nuclear weapons, with both sides pursuing nuclear war-fighting strategies, and with each fearing that the other is seeking to achieve a first-strike capability.

• The president's recognition of the danger in the present situation; his proposal to substitute a defensive strategy, based on a perfect defense, that would permit the destruction of all nuclear weapons; and the reasons why virtually all the experts consider such a goal unattainable.

• Alternative, "partial" defensive systems, which would be added to, not substituted for, offensive nuclear weapons, and which would almost certainly lead to a rapid escalation of the arms race and its extension into space.

• A totally different strategy, which would build on the ABM Treaty, move away from the nuclear-war-fighting mania, permit us to enter the twenty-first century with radically smaller nuclear forces (perhaps no more than 5 percent of the size of present inventories), and dramatically reduce the risk that our civilization will be destroyed by a nuclear conflagration.

• The way in which the Geneva negotiations can be structured to lay a foundation for a more secure tomorrow.

The superpowers' arsenals hold some 50,000 nuclear warheads. Each, on average, is far more destructive than the bomb that obliterated Hiroshima. Just one of our 36 strategic submarines has more firepower than man has shot against man throughout history. Thousands of nuclear weapons are ready for immediate use against targets close at hand or half a globe away, but just a few hundred warheads could utterly demolish the largest nation.

To deter war, each side seeks to persuade the other, and itself, that it is prepared to wage a nuclear war that would have the military objectives of a bygone age. What is known of Soviet nuclear-war plans is open to interpretation, but these plans appear to rely on tactics derived from Russia's pre-nuclear military experience. Current U.S. defense policy calls for nuclear forces that are sufficient to support a "controlled and protracted" nuclear war that could eliminate the Soviet leadership and that would even permit the United States to "prevail."

Nuclear-war-fighting notions lead to enormous target lists and huge forces. Our 11,000 strategic warheads are directed against some 5,000 targets. And NATO's war plans are based on early first use of some 6,000 tactical nuclear weapons in response to a Soviet conventional attack. Both NATO and the Warsaw Pact countries routinely train their forces for nuclear operations. War-fighting doctrines create a desire for increasingly sophisticated nuclear weapons which technology always promises to satisfy but never does. Today, both sides are committed to programs that will threaten a growing portion of the adversary's most vital military assets with increasingly swift destruction.

These armories and war plans are more than macabre symbols for bolstering self-confidence. Both Moscow and Washington presume that nuclear weapons are likely to be used should hostilities break out. But neither knows how to control the escalation that would almost certainly follow. No one can tell in advance what response any nuclear attack might bring. No one knows who will still be able to communicate with whom, or what will be left to say, or whether any message could possibly be believed.

When our secretary of defense, Caspar Weinberger, was asked whether it really would be possible to control forces and make calculated decisions amid the destruction and confusion of nuclear battle, he replied, "I just don't have any idea. I don't know that anybody has any idea." Surely it is reckless to stake a nation's survival on detailed plans for something about which no one has any idea.

It would be vastly more reckless to attempt a disarming first strike. Nevertheless, the arms race is driven by deep-seated fears held by each side that the other has, or is seeking, the ability to execute just such a strike.

The large force of powerful and increasingly accurate Soviet ICBMs has created the fear of a first strike in the minds of many U.S. leaders. According to this scenario, the Soviet missiles could, at one stroke, eliminate most of our Minuteman ICBMs; our surviving submarines and bombers would enable us only to retaliate against Soviet cities; but we would not do so because of our fear of a Soviet counterattack on our urban population; and thus we would have no choice but to yield to all Soviet demands.

A more subtle variant of this nightmare would have the Soviets exacting political blackmail by merely threatening such an attack.

Those who accept the first-strike scenario view the Soviet ICBMs and the men who command them as objects in a universe decoupled from the real world. They assume that Soviet leaders are confident that their highly complex systems, which have been tested only individually and in a controlled environment, would perform their myriad tasks in perfect harmony during the most cataclysmic battle in history; that our electronic eavesdropping satellites would detect no hint of the intricate preparations that such a strike would require; that we would not launch our missiles when the attack was detected; and that thousands of submarine-based and airborne warheads that would surely survive would not be used against a wide array of vulnerable Soviet military targets. Finally, they assume Soviet confidence that we would not use those vast surviving forces to retaliate against the Soviet population, even though tens of millions of Americans had been killed by the Soviet attack on our silos. Only madmen would contemplate such a gamble. Whatever else they may be , the leaders of the Soviet Union are not madmen.

That a first strike is not a rational Soviet option has also been stated by President Reagan's own Scowcroft Commission, which found that no combination of attacks from Soviet submarines and land-based ICBMs could catch our bombers on the ground as well as our Minutemen in their silos. In addition, our submarines at sea, which carry a substantial percentage of our strategic warheads, are invulnerable: in the race between techniques to hide submarines and those to find them, the fugitives have always been ahead and are widening their lead. As the chief of naval operations has said, the oceans are getting "more opaque" as we "learn more about them."

Despite all such facts, the war-fighting mania and the fear of a first strike are eroding confidence in deterrence. Though both sides are aware that a nuclear war that engaged even a small fraction of their arsenals would be an unparalleled disaster, each is vigorously developing and deploying new weapons systems that it will view as highly threatening when the opponent also acquires them. Thus our newest submarines will soon carry missiles accurate enough to destroy Soviet silos. When the Soviets follow suit, as they always do, their offshore submarines will for the first time pose a simultaneous threat to our command centers, bomber bases, and Minuteman ICBMs.

The absurd struggle to improve the ability to wage "the war that cannot be fought" has shaken confidence in the ability to avert that war. The conviction that we must change course is shared by groups and individuals as diverse as the freeze movement, the president, the Catholic bishops, the bulk of the nation's scientists, the president's chief arms-control negotiator, and ourselves. All are saying, directly or by implication, that nuclear warheads serve no military purpose whatsoever. They are not weapons. They are totally useless except to deter one's opponent from using his warheads. Beyond this point, the consensus dissolves, because the changes of direction being advocated follow from very different diagnoses of the predicament.

The president's approach has been to launch the Strategic Defense Initiative (SDI), a vast program for creating an impenetrable shield that would protect the entire nation against a missile attack and would therefore permit the destruction of all offensive nuclear weapons. The president and the secretary of defense remain convinced that this strategic revolution is at hand.

Virtually all others associated with the SDI now recognize that such a leakproof defense is so far in the future, if indeed it ever proves feasible, that it offers no solution to our present dilemma. They therefore advocate other forms of ballistic-missile defense. These alternative systems range from defense of hardened targets (for example, missile silos and command centers) to partial protection of our population.

For the sake of clarity, we will call these alternative programs Star Wars II, to distinguish them from the president's original proposal, which will be labeled Star Wars I. It is essential to understand that these two versions of Star Wars have diametrically opposite objectives. The president's program, if achieved, would substitute defensive for offensive forces. In contrast, Star Wars II systems have one characteristic in common: they would all require that we continue with offensive forces but add the defensive systems to them.

And that is what causes the problem. President Reagan, in a little-remembered sentence in the speech announcing his Strategic Defense Initiative on March 23, 1983, said, "If paired with offensive systems, [defensive systems] can be viewed as fostering an aggressive policy, and no one wants that." The president was concerned that the Soviets would regard a decision to supplement our offensive forces with defenses as an attempt to achieve a first-strike capability. That is exactly how they are interpreting our program; that is why they say there will be no agreement on offensive weapons until we give up Star Wars.

Before any further discussion of why Star Wars II will accelerate the arms race, it would be useful to examine why the president's original proposal, Star Wars I, will prove an unattainable dream in our lifetime.

The reason is clear. There is no evidence that any combination of the "defensive technologies" now on the most visionary of horizons can undo the revolution wrought by the invention of nuclear explosives. "War" is only one of the concepts whose meanings were changed forever at Hiroshima. "Defense" is another. Before Hiroshima, defense relied on attrition—exhausting an enemy's human, material, and moral resources. The Royal Air Force won the Battle of Britain by attaining a 10 percent attrition rate against the Nazi air force, because repeated attacks could not be sustained against such odds. The converse, a 90-percent-effective defense, could not preserve us against even one modest nuclear attack.

This example illustrates that strategic defense in the missile age is prodigiously difficult at best, an impression that is borne out by a detailed examination of all the schemes that propose to mount defenses in space. The term "defensive technologies" may conjure up images of mighty fortifications, but it refers to delicate instruments: huge mirrors of exquisite precision, ultrasensitive detectors of heat and radiation, optical systems that must find and aim at a one-foot target thousands of miles away and moving at four miles per second, and so forth. All these marvels must work near the theoretical limit of perfection; even small losses in precision would lead to unacceptably poor performance. Quite feeble blows against orbiting "battle stations" bearing such crown jewels of technology could render them useless.

Such attacks need not be surgical. If the Soviets were about to demolish us with a nuclear attack, they would surely not shrink from destroying our unmanned space platforms. And they have had nuclear-armed ABM interceptors ideally suited to that task for two decades. Such weapons could punch a large hole in our shield of space platforms, through which the Soviet first strike could immediately be launched. Hence any defense based on orbiting platforms is fatally vulnerable, or, as Edward Teller has put it, "Lasers in space won't fill the bill—they must be deployed in great numbers at terrible cost and could be destroyed in advance of an attack." The wide variety of countermeasures that have been developed during decades of ABM research show that every other proposed space-defense scheme has its own Achilles' heel.

The prospect of achieving the goal of Star Wars I has been succinctly put by Robert S. Cooper, the Pentagon's director of advanced research: "There is no combination of gold or platinum bullets that we see in our technology arsenal . . . that would make it possible to do away with our strategic offensive ICBM forces." Until there are inventions that have not even been imagined, a defense robust and cheap enough to replace deterrence will remain a pipe dream. Emotional appeals that defense is morally superior to deterrence are therefore "pernicious," as former Secretary of Defense James Schlesinger has said, because "in our lifetime, and that of our children, cities will be protected by the forbearance of those on the other side, or through effective deterrence." Harold Brown, also a former secretary of defense, has expressed the same thought.

Virtually everyone in the administration now agrees that a leakproof defense of population is not in the cards for decades, if ever. Therefore, while the president and the secretary of defense adhere to their original proposal, the technicians and others working on the SDI program are producing less radical rationales that blur crucial distinctions between hard-point defense, which is technically feasible, and comprehensive defense, which is not. These ever-shifting and intermingled rationales for Star Wars II call for careful scrutiny.

The most prominent fallback position is that even a partially effective defense would introduce a vital element of uncertainty into Soviet attack plans and would thereby enhance deterrence. This assumes that the Soviet military's sole concern is to attack us and that any uncertainty in their minds is therefore to our advantage. But any suspicions they may harbor about our wishing to achieve a first-strike capability would be inflamed by a partially effective defense.

Why? Because a leaky umbrella offers no protection in a downpour but is quite useful in a drizzle. That is, such a defense would collapse under a full-scale Soviet first strike but might cope adequately with the depleted Soviet forces that had survived a U.S. first strike.

Americans often find it incredible that the Soviets could suspect us of such monstrous intentions, especially since we did not attack them when we enjoyed overwhelming nuclear superiority.*

Nevertheless, the Russians distrust us deeply. They know that a first strike was not always excluded from U.S. strategic thinking, and they have never forgotten Hitler's surprise attack on them, in 1941, a disaster that dwarfed Pearl Harbor.

It would be foolhardy to dismiss as mere propaganda the Soviets' repeated warnings that a nationwide U.S. strategic defense is highly provocative. Their promise to respond with a large offensive buildup is no empty threat. Each superpower's highest priority has been a nuclear arsenal that can assuredly penetrate to its opponent's vital assets. No partially effective space defense can alter that priority.

Nor will those who now fear a Soviet first strike see their fears allayed by such a defense. On the contrary, these fears will be aggravated. The Soviet response will be based on traditional worst-case analysis, which will inevitably overestimate the effectiveness of our defense, just as in the 1960s and 1970s we targeted many more warheads on Moscow as soon as it was sur-

---

*In terms of numbers of strategic nuclear warheads, for example, the United States in 1960 had 6,300 to the Soviets' 200; in 1965 the figures were 5,000 to 600; in 1970, 4,500 to 1,800; in 1975, 8,000 to 2,700; in 1980, 9,200 to 6,000; in 1985, 11,100 to 8,500; and by 1990, assuming that U.S. and Soviet strategic forces are constrained by the SALT II agreement, the figures will be 13,600 to 13,000.

rounded by dubious ABM defenses. Being keenly aware of the fragility of our defenses, we would feel compelled to respond with a buildup of our own.

The claim that a Star Wars II defense would be a catalyst for arms reduction is therefore specious. Furthermore, arms control has been difficult enough when it has had to deal only with large offensive forces whose capabilities are relatively clear. It would be vastly harder to strike a bargain over space defenses whose effectiveness would be a deep mystery even to their owners, because they could never be tested under remotely realistic conditions.

Important support for Star Wars II stems from the belief that it best exploits our technological advantage in the inescapable competition with the Soviet Union. Those who hold this view ignore post-Hiroshima history and have less respect than we for the Soviet regime's ability to match our weapons and extract sacrifices from its people.

The U.S. invention of the atomic bomb was the most remarkable technical breakthrough in military history. And yet the Soviet Union, though devastated by war and operating from a technological base far weaker than ours, was able to create nuclear forces that gave it a plausible deterrent in an astonishingly short time. Virtually every technical initiative in the nuclear arms race has come from the United States, but the net result has been a steady erosion of American security. There is no evidence that space weapons will be an exception, for a crude nuclear blunderbuss can foil sophistication.

Then why are the Soviets so worried by Star Wars? Because strategic defense probably could succeed if the Russians played dead. For that reason they must respond. This will require vast expenditures they can ill afford, and will ultimately diminish their security. But that is equally true for us, whether we recognize it or not.

To summarize, these rationales for Star Wars II propose to achieve a superior strategic posture by combining unattainable technical goals with a policy rooted in concepts whose validity died at Hiroshima.

The public's intuitive awareness of the unacceptable risk posed by our present nuclear strategy is well founded. Our security demands that we replace that policy with one that is in firm touch with nuclear reality. If neither Star Wars I nor Star Wars II is the answer, what is?

The risk of catastrophic escalation of nuclear operations, and the futility of defense, lead us to base our proposal on the axiom that the initiation of nuclear warfare against a similarly armed opponent would be an irrational act. Hence, as we have said, nuclear weapons have only one purpose—that of preventing their use. They must not do less; they cannot do more. Thus, a restructuring of nuclear forces designed to reduce the risk of nuclear war must be our goal. All policies, every existing program, and each new initiative must be judged in that light.

Post-Hiroshima history has taught us three lessons that shape the present proposal. First, all our technological genius and economic prowess cannot make us secure if they leave the Soviet Union insecure: we can have either mutual security or mutual insecurity. Second, while profound differences and severe competition will surely continue to mark U.S.–Soviet relations, the nuclear-arms race is a burden to both sides, and it is in our mutual interest to rid ourselves of its menace. And third, no realistic scheme that would rid us of all nuclear weapons has ever been formulated.

The ultimate goal, therefore, should be a state of mutual deterrence at the lowest force levels consistent with stability. That requires invulnerable forces that could unquestionably respond to any attack and inflict unacceptable damage. If those forces are to remain limited, it is equally essential that they not threaten the opponent's deterrent. These factors would combine to produce a stable equilibrium in which the risk of nuclear war would be very remote.

This kind of deterrence posture should not be confused with the one currently prevailing among U.S. and Soviet nuclear forces. The 25,000 warheads that each nation possesses did not come about through any plan but simply descended on the world as a consequence of continuing technical innovations and the persistent failure to recognize that nuclear explosives are not weapons in any traditional sense.

The forces we propose could include a mix of submarines, bombers, and ICBMs. The land-based components should be made invulnerable in themselves, by some combination of mobile ICBMs and reductions in the number of warheads per missile. Two considerations would determine the ultimate size of the force: that it deter attack with confidence, and that any undetected or sudden violation of arms-control treaties would not imperil this deterrence. We believe that, ultimately, strategic forces having as few as 10 percent of the currently deployed warheads would meet these criteria, and tactical forces could be eliminated entirely. In short, the present inventory of 50,000 warheads could be cut to perhaps 2,000.

Before this goal is reached, other nuclear powers (China, France, Great Britain, and possibly others) will have to be involved in the process of reducing nuclear arsenals, lest their weapons disturb the strategic equilibrium.

The proposed changes in U.S. and Soviet strategic and tactical forces would require, as would the president's SDI, complementary changes in

NATO and Warsaw Pact conventional forces, or appropriate increases in NATO's conventional power. If the latter was necessary, it could be achieved at a fraction of the costs we will incur if we continue on our present course.

Having identified our goal, how can we move toward it? Some of our new policies would depend solely on the United States and its allies; others would require Soviet cooperation. The former should be governed by the dictum, attributed to President Eisenhower, that "we need what we need." Were we to drop futile war-fighting notions, we would see that many things we already have or are busily acquiring are either superfluous or downright dangerous to us, no matter what the Soviets do. Tactical nuclear weapons in Europe are a prime example, and the administration's policy of reducing their numbers should be accelerated. Other examples are programs that will haunt us when the Soviets copy them: sophisticated antisatellite weapons, sea-based cruise missiles, and highly accurate submarine-launched ballistic missiles. We are more dependent on satellites than the Soviets are, and more vulnerable to attack from the sea. Many of these weapons are valid bargaining chips because they threaten the Soviets, just as so much of their arsenal gratuitously threatens us.

Geneva provides an invaluable opportunity to take a giant step toward our goals. Is that not a preposterous assertion, the reader may well ask, for have we not claimed that Star Wars, which the president refuses to abandon, precludes arms control and guarantees an arms race? Surprisingly enough, it is not, if one takes account of a remarkable speech that Paul Nitze, the administration's senior adviser on arms control, gave in Philadelphia on February 20, 1985. If the points that Nitze made are accepted, it should be possible for the president to negotiate toward the goals we have set without abandoning a strategic-defense research program.

Nitze presented two criteria that must be met before the deployment of strategic defenses could be justified: the defense must work, even in the face of direct attack on itself, and it must be cheaper to augment the defense than the offense.

As we have seen, nothing that satisfies these criteria is on the horizon — a judgment in which Nitze apparently concurs, for he foresaw that during an initial period of "at least the next 10 years" no defenses would be deployed. During that period we would, in Nitze's words, "reverse the erosion" of the ABM Treaty. That is a window of opportunity, as we shall see.

Nitze envisioned the possibility of two additional periods following the first. In the second phase, some form of Star Wars II would be deployed alongside our offensive weapons, provided the two criteria he laid down had been met. If we entered the second phase, it probably would last for at least decades.

Ultimately, if Star Wars I proved practical, the second phase would be followed by a third, in which the leak-proof shield would be deployed and offensive weapons destroyed.

Nitze acknowledged that the problem of how to write an arms-control agreement during the second phase that would limit offensive arms while permitting defensive systems had not been solved. He said it would be "tricky." We agree. We know of no one who has suggested how to do it. But, by implication, Nitze was saying that this is an issue for future negotiations and that it need not stand in the way of a new agreement at this time.

Now back to the first phase, the window of opportunity. Why the fixation during this phase on the ABM Treaty? Because the treaty formalizes the insight that not just the deployment but even the development of strategic defenses would stimulate an offensive buildup. Were the treaty to collapse, we could not move toward our goal of reducing the offensive threat. Hence the fleeting window of opportunity: strengthening of the ABM Treaty coupled with negotiations on offensive strategic forces.

The treaty forbids certain types of radars and severely restricts the testing of components of ABM systems. Both of these provisions are endangered.

The Soviets are building a radar in Siberia that apparently will violate the ABM Treaty once it is completed. While this radar is of marginal military significance, it has great political import and poses an issue that must be resolved to the satisfaction of the U.S. government.

In the near future, the United States will be violating the restrictions on tests in spirit and probably in law if we place our research program on the schedule implied by Lieutenant General James Abrahamson, the director of the SDI, when he said, on March 15, 1985, that a "reasonably confident decision" on whether to build Star Wars could be made by the end of the decade or in the early 1990s. If we are unwilling to refrain from the tests associated with such a schedule, the Soviets will, with good reason, assume that we are preparing to deploy defenses. They will assiduously develop their response, and the prospect for offensive-arms agreements at Geneva will evaporate.The treaty's central purpose is to give each nation confidence that the other is not readying a sudden deployment of defenses: we must demonstrate that we will adhere to the treaty in that spirit.

The ABM Treaty does not forbid antisatellite weapons, and unless that loophole is closed we will have an arms race in space long before we have any further understanding of what, if anything, space defense could accomplish. Hence a verifiable ban on the testing of antisatellite weapons should become a part of the ABM Treaty regime. Because we are much more dependent on satellites than the Soviets are, such a ban would be very much in our interest.

A strengthened ABM Treaty would allow the Geneva negotiations to address the primary objective of offensive-arms control: increasing the stability of deterrence by eliminating the perceptions of both sides that the other has, or is seeking, a first-strike capability. This problem can be dealt with through hard-headed arms control. There is no need to rely on the adversary's intentions: his capabilities are visible. Mutual and verifiable reductions in the ratio of each side's accurate warheads to the number of the other side's vulnerable missile launchers could reduce the first-strike threat to the point at which it would be patently incredible to everyone. Both sides have such immense forces that they should concentrate on quickly reducing the most threatening components—those that stand in the way of stability and much lower force levels.

What is needed is deep cuts in the number of warheads, but cuts shaped to eliminate the fear of first strikes. Because the two sides have such dissimilar strategic forces, this process will be very difficult, but it should be possible in the first phase to accomplish reductions of 50 percent. It would be reasonable, for example, for the United States to insist on large reductions in the number of Soviet ICBM warheads, but in the bargaining we must be ready to make substantial cuts in our counterpart forces, including, for example, the silo-killing submarine-based D-5 missile.

In sum, the arms negotiations now beginning in Geneva represent a historic opportunity to lay the foundation for entering the twenty-first century with a totally different nuclear strategy, one of mutual security instead of war-fighting; with vastly smaller nuclear forces, perhaps 2,000 weapons in place of 50,000; and with a dramatically lower risk that our civilization will be destroyed by nuclear war.

Several themes should govern our attitude and policies as we move through those negotiations toward our long-term objectives.

Each side must recognize that neither will permit the other to achieve a meaningful superiority; attempts to gain such an advantage are dangerous as well as futile.

The forces pushing each side in the direction of a "first-strike" posture must, at least from the standpoint of the adversary, be reversed. A stable balance at the lowest possible level should be the goal.

Our technological edge should be exploited vigorously to enhance our security, but in a manner that does not threaten the stability of deterrence. Space surveillance and data processing, which form a large portion of the SDI program, illustrate what technology could contribute to treaty verification.

We must not forget Winston Churchill's warning that "the Stone Age may return on the gleaming wings of science," and we must learn to shed the fatalistic belief that new technologies, no matter how threatening, cannot be stopped. While laboratory research cannot be constrained by verifiable agreement, technology itself provides increasingly powerful tools that can be used to impede development and to stop deployment. For example, only an absence of political will hinders a verifiable agreement preventing the deployment of more-threatening ballistic missiles, because they require many observable flight tests.

We must also allay legitimate fears on both sides: the Soviets' fear of our technology, and our fear of their obsessive secrecy. These apprehensions provide an opportunity for a bargain: Soviet acceptance of more-intrusive verification in return for American constraints on applications of its technological innovation. Penetration of Soviet secrecy is to our mutual advantage, even if the Kremlin does not yet understand that. So is technological restraint, even though it runs against the American grain.

We have reached our absurd confrontation by a long series of steps, many of which seemed to be rational in their time. Step by step we can undo much of the damage. The program sketched in this article would initiate that process. It draws on traditional American virtues: striving with persistence and resourcefulness toward a high but attainable goal.

This program would steadily reduce the risks we now face and would begin to restore confidence in the future. It does not pretend to rid us totally of the nuclear menace. It addresses our first duty and obligation: to assure the survival of our civilization. Our descendants could then grapple with the problem that no one yet knows how to attack.

# Chop Down Nuclear Arsenals

fter the unpleasant presidential campaign, it is hard to tell what
George Bush will do as president. But I take heart from a *New
Yorker* editor's comment after the 1940 election: "Most of us
speak better when not running."

As a candidate, Bush often declared that he represented the mainstream.
Polls indicate that the mainstream wants arms control: more than 80 percent of the respondents in an October 1988 survey by Daniel Yankelovich
supported a bilateral freeze on nuclear weapons. Some 80 percent of the
respondents in a *Wall Street Journal* poll—including those who called
themselves conservative—believed military spending should either
decrease or stay constant. So I take the optimistic view that President Bush
will build on Ronald Reagan's arms control initiatives and will pursue a 50
percent reduction in strategic nuclear weapons.

There are problems with the strategic arms reduction talks (START),
some real and some imaginary. The real problems arise from the question
of how U.S. strategic weapons should be structured following an agreement; the imaginary ones arise from feeling rather than analysis, as I will
discuss later. Ultimately, I believe, START treaties can clear the way for a
very large reduction in nuclear weapons.

My optimism derives in a large part from the tremendous changes that
have been occurring in the Soviet Union. The leader in these
changes is Mikhail Gorbachev. But Soviet intellectuals have apparently
realized for a long time that change was necessary, chiefly because the
Soviet economy simply did not work. By last fall, evidence of change was
rapidly accumulating:

• On October 4, 1988, Vadim A. Medvedev, the new chief ideologist,
announced a major reversal in Soviet thought. In a speech to political scientists from communist countries, he said: "Present-day realities mean that

universal values such as avoiding war and ecological catastrophe must out-weigh the idea of struggle between classes." And he added another signal that the Soviet Union would become easier to live with: "Peaceful coexist-ence is a lengthy, long-term process whose historic limits are difficult to determine."

Medvedev, an economist, called for major decentralization of the econ-omy. A socialist country, he said, must learn from other socialist countries and even from the capitalist West.

• On October 14, 1988, Gorbachev proposed that farms be leased back to farmers who would pay a fixed rent to the government and keep the remainder of the proceeds for themselves.

• Andrei Sakharov, who had earlier been permitted to return to Moscow from his Gorky exile, was made a member of the Presidium of the Soviet Academy of Sciences in October 1988. In November, he was allowed to travel to the United States.

• The Soviets' tremendous concessions on intrusive arms control inspections made possible the treaty on intermediate-range nuclear forces and by fall, missiles were being destroyed in the presence of witnesses from both sides. Inspectors have been stationed permanently on the periph-ery of factories that formerly manufactured INF missiles. Soviet inspectors in Utah were reportedly impressed by the friendliness of local citizens.

In the same spirit of friendly cooperation, nuclear weapons tests in Nevada and Semipalatinsk were observed and measured jointly by the host country and guest scientists. These observations made it possible to com-pare the equipment used to measure the yield of weapons tests.

• Finally, in a dramatic speech to the United Nations on December 7, 1988, Gorbachev promised major unilateral reductions in Soviet conven-tional forces. He offered to reduce Soviet armed forces by half a million men and 10,000 tanks and to cut several other important weapons. While this does not make Soviet forces equal to NATO's, at least in quantity, it is a very good and generous start.

Gorbachev's peaceful revolution and pragmatism were chiefly responsi-ble for the INF Treaty and the good beginning in START, although the U.S. policy of "negotiating from strength" may have helped reveal to the Sovi-ets the futility of continuing the arms competition, just as their economic policy had entered a blind alley.

But Gorbachev and the reform party may lose power. It remains to be seen whether the domestic changes will work, and until the ordinary Soviet citizen feels that life has improved, the reforms will remain fragile. One can only hope that the transfer of farms to private initiative will improve food production, and that the loans from Western Europe will improve the production of consumer goods.

Some analysts recommend that the United States proceed with caution in arms control negotiations until Gorbachev's future is more certain. I believe, on the contrary, that this is a time of extraordinary opportunity, and that we must use it. If Gorbachev and company were, unhappily, to be replaced by a hardline regime, we would all be better off with an agreement in place that reduced the absurd numbers of strategic weapons. An agreement that permitted additional intrusive inspection would be even better. It would be a big step toward an open world and a complete break with the policy of secrecy for which Tsarist Russia was already noted.

The Soviet government has recognized, as the Western arms control community did long ago, that neither side can be safe unless both are safe. And the Soviets now want arms control more urgently than the U.S. government does.

When the Los Alamos laboratory had built the first nuclear weapons, we scientists thought that if national arsenals ever included them at all there would be very few, perhaps a few dozen. Now the numbers have increased a thousandfold, so that the 50 percent cut envisaged in START is only a small step back toward the original concept.

But while their numbers are being reduced, the weapons should be made more survivable, so that there is less incentive to use them in a preemptive strike. Submarines are the best choice for a survivable U.S. nuclear force, because they are difficult to detect. Fortunately for strategic stability, it is highly unlikely that effective means can be developed to detect and attack deeply submerged missile-carrying submarines. Even if antisubmarine warfare were to become possible someday, it will not happen overnight. There would be ample time to shift emphasis to the land and air legs of the triad.

So the most important genuine objection to a START agreement has been raised by the U.S. Navy: that after the reductions there will be too few missile-carrying submarines. I agree with the navy that the United States needs many such submarines. The security of the U.S. deterrent depends on the difficulty of detecting submarines, and the difficulty increases with the number at sea. But there is no law that says each submarine must carry nearly 200 nuclear warheads. If the force must be cut in half, the same number of submarines should carry half the number of warheads. For now, as Richard Garwin has suggested, half the launch tubes could be filled with concrete; in the future, smaller submarines should probably be built. And it does not matter whether submarine missiles carry single or multiple warheads.

Another stumbling block is ship-based cruise missiles. It would be best to eliminate nonstrategic nuclear weapons on ships altogether, as Reagan's arms control advisor Paul Nitze proposed. But this leads to difficult verification problems. Perhaps the best we can do is limit the number and take it on faith that both sides will obey the agreement. Sidney Drell of Stanford University has suggested that this might be acceptable for sea-launched cruise missiles because they are rather slow and thus not first-strike weapons.

With submarines secure for the present, one may question the need for land-based intercontinental ballistic missiles (ICBMs). For a long time ICBMs were the most accurate missiles, but with the introduction of the accurate D-5 (Trident II) submarine-launched missile, there is no longer a good reason to emphasize ICBMs. Land-based missiles are needed only as insurance against unforeseen advances in antisubmarine warfare, or against failure of communication with submarines. But what kind of ICBMs should the United States deploy?

• *MIRVs*. Even Henry Kissinger, who approved of the decision to introduce MIRVs (multiple independently targetable reentry vehicles) around 1970, now admits that they must be eliminated on land-based missiles. MIRVed missiles are first-strike weapons, threatening to the enemy and inviting preemptive attack. In the 50 percent reduction, both sides should move decisively toward single-warhead ICBMs.

• *Mobility*. The Soviet Union has an enormous land mass, lots of open space, and few coasts on the open oceans. The United States has long coasts and is more densely populated. To have a survivable force, the Soviets need mobility on land, while the United States does not. I like Drell's idea for improving the survivability of the U.S. land force: to put small, single-warhead missiles ("Midgetman") in silos, for the time being. They would be unattractive targets, because the Soviets would have to use two warheads to destroy each missile, and they would still be exposed to the much more numerous U.S. submarine missile force. Putting Midgetman in silos would save a great deal of money. The missiles could be designed in such a way that they could be put on an all-terrain vehicle if this were ever to become necessary. A force of, say 400 small silo-based missiles would bring us closer to an ultimately desirable force than would the ten-warhead MX in a rail-garrison mobile basing system.

• *Yield*. Now that both sides have achieved great accuracy, heavy missiles such as the Soviet SS-18 are no longer necessary, and all ballistic missiles should be limited in yield. The Soviets have promised to reduce numbers of SS-18s by half and also to reduce the total throw weight of their missile force by half. U.S. Minuteman II and III missiles are much smaller, but their yield should be further reduced, in line with their increased accuracy. One reason for such reduction is to make sure that there will not be a "nuclear winter" if the missiles are ever launched: whether the temperature falls by four degrees or twenty, it would be a global catastrophe.

Congress has supported the move toward smaller missiles. But the Midgetman has been conceived as a mobile missile, which is unnecessary. And the air force has proposed putting three warheads on each missile instead of one, in order to save money. This would defeat its purpose. Survivability, not cost, should be the decisive factor. Arms reduction will not necessarily save much money at first, but savings will come later with reduced maintenance costs.

Once START is concluded, other important arms control initiatives can follow:

• Restricting the number of permissible missile tests would be a powerful measure, although it is seldom mentioned. Missile tests are above ground and are therefore easy to verify by satellites and other "national technical means." Limiting missile tests would more effectively halt the development of new weapons than would a comprehensive nuclear test ban.

• Tactical nuclear weapons, with a range less than 500 kilometers, should be reduced: each side now has thousands. If war broke out in Europe, the U.S. doctrine of flexible response envisages possible escalation to tactical nuclear war. This would be extremely dangerous for Western Europe, especially since nuclear artillery, a large part of the tactical force, is not very mobile and may present the commander with the well-known dilemma, "use them or lose them." And tactical nuclear war may quickly lead to a full, strategic nuclear war.

• During the long period that will be required to implement START, negotiations should begin to reduce strategic weapons further, reasonable goal is another 50 percent reduction, to 3,000 warheads on each side. MIRVs must continue to be eliminated and survivability must be emphasized.

• At this stage, China, France, and Great Britain must be brought into the negotiations. This will not be easy: France may be the most resistant, and there will presumably be long haggling about the number of warheads permitted to each country. Other suspected weapons countries, such as India and Israel, will need to join the third round.

The ultimate goal should not be zero nuclear weapons but some low number like 200 to 1,000 on each side. Fortuitously, the Committee of Soviet Scientists for Peace has come up with the similar number, 600. One reason for retaining some weapons is residual mistrust between the superpowers which requires a hedge against possible treaty violations. Dozens of missiles may be concealed, but it would be very difficult to conceal hundreds, ready to launch. A second reason is that rogue countries might start small nuclear arsenals. Finally, nuclear weapon design is well known, and there is little difference between having the design and having a small residual force.

Among the imaginary problems, the most important is the Strategic Defense Initiative. Star Wars is the misbegotten product of an uncritical president and overenthusiastic advisers.

An informed decision on whether such a system can be deployed is at least a decade away, according to a 1987 study by a panel of the American Physical Society who had access to secret laboratories of the SDI Organization. Current SDI devices perform at one-hundredth to one-ten-thousandth the level that would be required. In 1988, a panel of the General Accounting Office came to similar conclusions.

Everyone, including SDIO Director James D. Abrahamson, agrees that an "astrodome" defense that would protect everyone in the United States is impossible. It may be possible to defend valuable military assets if they are hardened, that is, if they can survive very high pressures. But ICBM silos, which can be hardened, are not really worth defending, but it is doubtful that they can be made hard enough. In any case, such "point" defense does not require space weapons.

Many who are on the fence about SDI believe that vigorous research should continue. A 1986 meeting of the Aspen Strategy Group endorsed further research but strongly recommended focusing on technology development in the laboratory, not spectacular demonstrations in space. This group included General Brent Scowcroft, who will direct the National Security Council in the Bush administration.

This approach would make it possible to stay within the traditional interpretation of the 1972 Anti-Ballistic Missile Treaty, rather than the broad interpretation favored by President Reagan—according to which exotic space weapons should be permitted because nobody thought of them in 1972 and therefore they are not explicitly forbidden. Georgia Democratic Senator Sam Nunn has shown by a convincing legal argument that only the traditional interpretation is justified.

If the United States abandons the traditional interpretation there is great danger that the Soviets would back out of the treaty in short order and deploy a ground-based antiballistic missile system. General Robert T. Herres, vice chairman of the Joint Chiefs of Staff, recently told the congressional armed services committees that the ABM Treaty, traditionally interpreted, is essential for U.S. security.

Some people favor a minimal strategic defense against accidentally launched missiles or missiles from a third country, something like the anti-China ballistic missile defense system that was proposed in 1967. But a small-scale SDI would not be worth the cost, and it would be the camel's nose in the tent. A better way to insure against accidental launch would be an electronic disabling device activated after launch. And a rogue country could presumably deliver nuclear weapons in other ways, for example, by bringing a ship into New York harbor.

A modest SDI is undesirable for another reason: it could be an effective antisatellite weapon. Because our security depends on a well-functioning system of intelligence satellites, we should promptly conclude a treaty prohibiting antisatellite weapons instead of pursuing SDI.

SDI proponents say defense is more moral than offense. This may be so, but there is no effective defense against nuclear weapons. Reagan's goal of making nuclear weapons "impotent and obsolete" can be achieved only by negotiation, not by technical devices. SDI may have helped bring the Soviets to the negotiating table, as some conservatives claim. But the main reason was the fundamental change in Soviet policy under Gorbachev. And now that they are at the table, SDI is no longer necessary for this purpose.

Defense against the present large Soviet missile force would involve exorbitant expenditure and would still be ineffective. It may be possible to defend against a very small force of perhaps 200 strategic nuclear warheads, but that would be unnecessary and destabilizing.

Another imaginary problem has been posed by U.S. Congressman Les Aspin, the Wisconsin Democrat who chairs the House Armed Services Committee and who generally is rather pragmatic. Aspin worries about warehoused Soviet missiles whose numbers START would not restrict. In case of a nuclear war, it would take a long time to bring such missiles onto a launch pad, and by that time the war would probably be over. And if the Soviets are willing to permit on-site inspection, they may well agree to declare the number and location of such missiles.

Security for both sides ultimately depends on freedom from fear of attack. At present the main Western concern is the apparent superiority of the Warsaw Pact over NATO in conventional weapons. The perpetual Soviet concern is invasion from the West, repeating that of the Nazis in

1941–1943, the Germans in 1914–1917, and Napoleon in 1812.

Gorbachev began addressing the Western concerns in his December 1988 speech to the United Nations. And the problems will soon be discussed in the Vienna talks on conventional forces, which will replace the talks on Mutual and Balanced Force Reduction. Unfortunately, NATO members disagree on what the goals of the talks should be. But in the United States there is not much internal dispute: both political parties want to reduce conventional forces in Europe.

An important first step has already been achieved in the agreement signed by 35 nations in Stockholm in 1986, providing for notification and observation of maneuvers. This builds confidence that neither side will be able to launch a massive surprise attack with conventional weapons. And the concepts of nonoffensive defense spelled out by West European analysts and taken up by the Soviets raise encouraging possibilities but some questions as well—about the role of planes, for instance. The initial data exchanges agreed to in the May 1988 Moscow summit will be important, since there has been persistent disagreement on the numbers of weapons each side possesses. And the West will have to decide what it wants to give in exchange for reductions in Soviet tanks, motorized artillery, and perhaps interceptor planes. Perhaps NATO might agree to reduce fighter bombers like the F-16.

Reconciling the divergent European views will require strong leadership on the part of the U.S. president. Even so, the talks will be difficult, and it would be fatal to hold START hostage to success in conventional force negotiations.


A fundamental change in the attitude and character of a country—or *Bewussteinswandel*, to use the title of a recent book by Carl Friedrich von Weizsäcker—is possible: we have seen it in Germany's transformation from Nazi empire to federal republic. This is happening today in the Soviet Union, fortunately without a war.

But the West must undergo its own *Bewussteinswandel* and realize that the Soviet government has changed and is no longer an "evil empire." We may pursue and stabilize this mutual, basic attitude change in joint enterprises—environmental problems, especially the control of carbon dioxide and ozone, should be top priorities—and in agreements not to excessively arm opposing parties in local conflicts. If "avoiding war and ecological catastrophe" really does outweigh the idea of a class struggle in the Soviet Union, then peaceful coexistence is possible.

# THE
# FREEZE

# The Value of a Freeze

## WITH FRANKLIN A. LONG

The rapid increase in public support for a nuclear-freeze agreement — that is, a mutual freeze on the testing, production and further deployment of nuclear weapons — has been a remarkable political phenomenon. In less than a year, support has grown from a few volunteers collecting signatures on petitions to a Congressional vote in which supporters of a freeze very nearly prevailed. In the fall of 1982, eight states and the District of Columbia will vote on freeze referendums. Already Wisconsin voters have overwhelmingly voted yes in such a referendum.

There are many reasons for this strong support for a freeze, including fear of nuclear war, resistance to high levels of military spending and opposition to particular military policies of the Reagan administration. But to most supporters, the chief purpose of a freeze is simple: It is to help stop an immense, continuing, dangerous and incredibly costly arms race between the two superpowers.

The administration opposes a prompt freeze. Its members offer a variety of arguments why a freeze is a bad idea. Most of these arguments lack validity.

One argument that spokesmen offer is that a freeze would leave no incentives for the Soviet Union to stop the arms race. But strong incentives to stop it already exist in both countries. The Soviet Union shares with us an unbalanced economy caused by immense expenditures for military systems. A freeze that permitted large decreases in military spending would be of great help to the civilian economies of both countries.

Another argument is that America would be "behind" the Soviet Union if a freeze were agreed upon. This is highly debatable. American military leaders have argued that our nuclear forces are preferable to the Soviet

Union's, and even our more pessimistic military leaders agree that there exists an approximate nuclear parity and a situation of strong mutual strategic deterrence.

The administration complains that in the 1970s, the Russians built up their nuclear arsenal relentlessly while America stood still. In fact, the number of warheads in our strategic forces increased from about 4,000 in 1970 to 10,000 in 1980, while the Soviet Union's increased from about 1,800 to 6,000 in 1980 and 8,000 in 1982. The Soviet buildup followed ours by about five years. The best way to stop still further buildups is a freeze followed by negotiated, substantial arms reductions.

One administration spokesman bases some of his arguments against a freeze on a proposition with which we agree—namely, nuclear weapons "are good if they promote stability and contribute to deterrence of war, and bad if they diminish stability and weaken deterrence." But then he argues in favor of all components of the administration's nuclear arms buildup, whether they lead to stability or not.

Consider two new American delivery systems: the proposed intercontinental ballistic missile called the MX, and the planned deployment of highly accurate cruise missiles on submarines. Both weapons, if deployed, will be seriously destabilizing. All plausible arrangements for basing the MX will leave it vulnerable to Soviet attack; moreover, the threat to Soviet ICBMs from the high accuracy of the MX is an added reason for the Russians to launch a nuclear first strike with their own ICBMs. Our submarine-launched cruise missile will be destabilizing because of the serious difficulty, in reaching arms-control agreements, in verifying the numbers that are deployed.

It will take statesmanship and a mutual desire for peace to negotiate a freeze. Either country can obstruct the negotiations by unrealistic conditions or by demands for excessively intrusive verification procedures. But verification need not be a severe problem, since both countries have substantial national technical means for verification. Furthermore, it is clearly easier to verify zero activity—that is, no testing, no production, no deployment of new systems—than to verify quotas or restrictions.

The larger goal for Washington and Moscow is to obtain some measure of political reconciliation, based on a mutual understanding, that neither party benefits from the current costly and dangerous confrontation. Arms-control agreements will still be needed to reduce the world's arsenals of nuclear weapons. The Strategic Arms Reduction Talks (START) negotiations and those on reductions in intermediate-range nuclear forces under way in Geneva should continue. Both sides still need other political agreements and confidence-building measures. Both must work to decrease greatly the threat of major war in Europe. But for all of these, a mutually agreed nuclear freeze would be an important first step, a clear signal for new directions.

# Debate: Bethe vs. Teller

## RESPONSE BY EDWARD TELLER

The call for a bilateral nuclear-weapons freeze, the subject of Proposition 12 before California voters on November 2, 1982 is meant to be a signal to the U.S. government that the American people want the arms race to stop.

I was one of the scientists at the Los Alamos National Laboratory, which developed the atomic bomb during World War II. We thought at the time that the United States might deploy a few dozen nuclear weapons. Not in our worst nightmares did we imagine that someday there would be about 10,000 strategic nuclear weapons in the United States and a similar number in the Soviet Union. These large numbers make no sense, even if we wished to destroy *all* military and industrial targets in the Soviet Union. At this level of armament, as Henry Kissinger said many years ago, it is meaningless to ask who is ahead and who is behind.

President Reagan, when he asked for large reductions in the strategic nuclear force by both sides and authorized negotiations in Geneva, recognized that these large numbers of weapons make no sense and give no security to either country.

Yet, at the same time, the United States is engaged in a massive buildup of nuclear armaments. The supporters of this arms buildup claim that the freeze would put us into a position of permanent inferiority.

What is the purpose of strategic nuclear forces? There is really only one: to deter a nuclear attack by a potential enemy. For this purpose you need to be able to retain the capacity to inflict unbearable damage on the enemy even after he has inflicted such damage on you. To accomplish this we have placed more than half of our nuclear warheads on submarines, which are virtually undetectable in the ocean. Therefore, missiles and submarines are essentially invulnerable.

It is true that the Soviets have more and bigger land-based intercontinental ballistic missiles. How could it be different? Overall the two countries have approximate parity in nuclear weapons. We have put our emphasis on submarine-launched missiles, so obviously we have fewer land-based missiles. Some strategic analysts have expressed fear that with their large ICBMs of increasing accuracy, the Soviets might be able to destroy our land-based ICBMs by targeting their own missiles against our missile silos. But this would not incapacitate our ability to retaliate: We would retain the major part of our deterrent force, the submarines.

Moreover, if ICBMs are becoming vulnerable, then the Soviets, having three-quarters of their weapons in ICBMs, will be more vulnerable than the United States. Only about one-quarter of Soviet nuclear warheads are on submarines. Moreover, the fraction of Soviet submarines actually deployed at sea is considerably smaller than ours, partly for geographic reasons, and partly because Soviet submarines need more frequent repairs.

In any case, our submarine force is sufficient to deter any massive nuclear attack by the Soviets on the United States.

The arms race must stop. At every step, we have taken the lead. We were the first to deploy ICBMs by the hundreds. We invented MIRV, the multiple independently targeted reentry vehicle, by which one missile can send many, in some cases 10, warheads to different places in enemy territory. Despite a warning from our Arms Control and Disarmament Agency that MIRV would benefit the Soviets, with their heavier ICBMs, more than us, we refused to include it in the SALT I agreement. We deployed MIRV; the Soviet Union followed a few years later, with the result that these are the Soviet missiles our Defense Department now fears the most. If we engage in another arms buildup, it will again cause a Soviet countermove.

Arms-control treaties have been verified without on-site inspection, using especially intelligent satellites. They are our eyes that can distinguish objects as small as an automobile. The Soviet Union has a similar capacity to see what goes on in the United States. Intelligence satellites are the silent communication systems between countries. Because of them, each country can monitor ICBM emplacements, submarines in docks, troop movements, or any kind of unusual activity that might be a warning. Without them we are blind, and fear and distrust increase. It is essential that we preserve them. Therefore, it is deplorable that the U.S. Air Force is beginning tests of antisatellite weapons, which are to go into service in 1986. Assuming that the Soviets will follow our example, as they have always done, we are setting the stage for blinding ourselves deliberately. There should be a treaty prohibiting the testing and deployment of antisatellite weapons.

One of the worst features of the arms race is that nationalistic passions are excited in order to make Congress and the taxpayers willing to finance the arms. These passions in turn make war more likely. Instead, the goals for Washington and Moscow ought to be some measure of political reconciliation, based on a mutual understanding that neither party benefits from the current costly and dangerous confrontations.

We must give a signal to the administration and to Congress that we want the arms race to stop. Proposition 12 is such a signal.

---

# EDWARD TELLER RESPONDS

From 1957 to the present, various U.S. administrations have engaged in almost continuous negotiations with the Soviet Union on arms limitation. What has been the net result?

In 1957, the United States was the strongest military power and the world seemed safe from immediate danger of holocaust. This year President Reagan had the honesty and courage to tell the American people that the Soviet Union is ahead of us in nuclear arms. The "balance of terror" was never an attractive expression, but today the terror is clear to all of us and the balance seems to be weighted against the free world.

The supporters of Proposition 12, the nuclear-freeze initiative on California ballots November 2, 1982 do not claim that it contains a new idea. They only claim their proposition is "as simple as a can opener." Anyone can understand it. Oversimplifying the world's most important urgent problem is possible, but doing so solves nothing.

Everyone agrees that there can be no more important goal than avoiding the catastrophe of a third world war. Though the best scientific evidence, including a 1975 report of the National Academy of Sciences, assures us that the human race would survive a nuclear conflict, such a holocaust would almost certainly claim many more victims than the 50 million killed by Adolf Hitler.

The proponents of the freeze claim they want to have a bilateral arrangement. Indeed, the Soviet government endorses the idea. Yet, while in New York 750,000 people demonstrated in favor of the great "can opener" to strains of rock music, in Moscow, 11 would-be demonstrators went straight to the Gulag Archipelago.

Even if a freeze agreement was signed, sealed and delivered, what would it mean? The United Nations agreement prohibiting the manufacture, stockpiling and use of biological and chemical weapons has been signed by many nations, including the Soviet Union. A U.S. State Department report (No. 98, dated March 22, 1982) details evidence that more than 11,000 people in Laos, Cambodia and Afghanistan have been killed by toxic weapons supplied by the Soviet Union. United Nations teams have been denied admission for on-site inspections.

How much power does an agreement have to deter? Has world public opinion been effective in preventing violations involving horrible chemical and biological weapons? Nuclear weapons are immensely destructive, but their effects remain limited. Damage from self-multiplying organisms is much harder to predict or limit.

Would it not be better, if instead of relying on the freeze, we relied on true defensive weapons? Several ingenious possibilities exist: Incoming nuclear missiles, for example, could be harmlessly destroyed.

The Soviet leaders want power and more power, but they don't want to take risks. Protective defense and civil defense, any action which makes their success questionable, serves as a deterrent. This kind of deterrence is incomparably more humane than deterrence by Mutual Assured Destruction (MAD).

As the advocates of a freeze implicitly demand, it is high time that we turn away from the MAD concept. But in actual fact, the freeze will not accomplish this aim but will make its accomplishment more difficult. A return of the military establishment to its elementary purpose—to defend (in the most literal sense of the word) the American people—is morally more justified and much more realistic.

Such defense may be carried out by non-nuclear or nuclear weapons. We should use whatever is needed—whatever will assure survival while hurting the fewest number of people, even on the enemy side. Whatever will accomplish this aim with the greatest and earliest probability of success is the obvious choice.

But what are these miraculous weapons that can defeat a rocket approaching with a velocity greater than that of a bullet? The American public is ignorant, even of the general ideas on which they are based. Yet it is most unlikely that the Soviet leaders don't know.

James Madison commented long ago that a democracy without free information is a farce, a tragedy, or both. The advocates of the freeze are orchestrating this farce. Their followers are the players in a tragedy. The final result may be worse than any that we can imagine.

A vote for the freeze will not only deflect people's minds from the obvious need for defense but will make work on protective defense much more difficult, if not impossible. A vote against the freeze is a statement of the overriding and lasting necessity of being able to defend oneself. This vote should be joined by a ground-swell movement to convert our policy to one of true defensiveness, for only in this manner can we hope to achieve a lasting stable peace.

Everyone knows of the horrors of World War II. Some even remember the days preceding it when the British prime minister, Neville Chamberlain, went to Munich, negotiated with Hitler, and brought home "peace in our time." That peace lasted one year.

The freeze advocates are Chamberlain's worthy successors. Chamberlain was no Nazi, just as the supporters of a freeze are not communists. They are frightened, uninformed, unreasoning people. Fear is a poor counselor. Their good intentions could take us all down a most horrible road.

Peace is not the result of a mere wish. It must be built using reason and hard work. Defensive weapons, improved cooperation with our allies and more completely informed public opinion are all-important building blocks. By defeating Proposition 12, we open the road of constructive steps leading to peace.

# After the Freeze Referendum

## WITH FRANKLIN A. LONG

T he voting on nuclear freeze resolutions that took place during the U.S. national elections in November 1982 demonstrated wide citizen approval for a freeze. By a strong margin, voters in states that contain over a quarter of the people in our country supported a freeze by the United States and the Soviet Union on deployment of strategic nuclear weapons. Only in one state, Arizona, did voters reject a freeze resolution, and there by a narrow margin. The wording of the freeze resolution differed from state to state, but the common message was plain: "Let's put a stop to the nuclear arms race."

We strongly hope that this message has reached the White House and that greatly increased efforts to negotiate nuclear arms reductions with the Soviet Union will be one result. Negotiations are underway in Geneva, but the pace is slow, the positions of the two sides are far apart, and there is no visible sense of urgency.

Fortunately, there is an immediate step that the United States by itself can take to advance these negotiations: the U.S. Senate can ratify the SALT II treaty. Here is a treaty containing important constraints on nuclear weapons that has already been negotiated and signed by the United States and the Soviet Union, and ratified by the latter nation. Its principal provisions are being adhered to by both nations. By ratifying this treaty, the present U.S. administration can formalize its commitment to these provisions and also signal the seriousness of its intention to negotiate further arms reductions.

But what about a nuclear freeze? In our judgment, even if SALT II is ratified by the United States—and especially if it is not ratified—there are strong arguments for quickly negotiating a bilateral freeze agreement.

If a nuclear freeze is to be negotiated soon, and in a way that does not interfere with vigorous negotiations on arms reductions, the treaty should be uncomplicated, and the provisions for verification should be simple and straightforward. An illustration of a first-stage treaty that we believe would be negotiable and be an important step forward is a *freeze on the testing and deployment of all long-range nuclear weapons systems*, for example, with maximum ranges over 1,000 kilometers. This freeze could be verified by national technical means, that is, with each nation monitoring the other by well-established satellite observation and related procedures. It might be necessary to make provisions for a small annual quota of "proof tests" of current weapons systems to help maintain readiness. The bilateral Standing Consultative Committee established to monitor the SALT I agreements could be used to resolve any compliance ambiguities, and a provision might be included to review the agreement every five years as with the 1972 Antiballistic Missile Treaty.

The other substantive constraints envisaged in a comprehensive freeze —a ban on the production of weapons-grade fissile materials and on new nuclear warheads, and a comprehensive ban on the testing of nuclear explosive devices—could be negotiated in a second stage. All these provisions are important in stopping the nuclear arms competition, but we believe a two-stage approach will facilitate the negotiation of acceptable verification procedures for these additional components. Our emphasis on a simple first-stage treaty stems from a strong desire to see negotiation and ratification of a freeze treaty accomplished within a 12-month period, something that seems feasible if treaty provisions are straightforward and mutually acceptable, and if negotiations proceed at a vigorous pace.

There are, of course, other proposals for a nuclear freeze that have been suggested, and that might also serve as a negotiable first stage. The SALT II treaty itself provided for several agreed "limits," which, in effect, are freezes, on numbers of delivery systems, and on numbers and types of MIRVed systems. Still other freeze proposals were formally presented by the United States, one as early as 1964. Given this history of study and consideration, it should be possible for both the United States and the Soviet Union to move quickly to formal negotiations for a mutually acceptable first-stage nuclear freeze.

Experience with other agreed U.S.–Soviet freezes augurs well for a deployment and testing freeze that will be obeyed by both parties. During the 25 years that the two nations have been negotiating on arms control there have been two periods of freeze, or more precisely, one period of a joint voluntary moratorium on tests of nuclear explosive devices and one formal freeze on launchers for nuclear-armed ballistic missiles. Both were important in themselves, and helpful in permitting serious negotiations to go ahead. The 1958 to 1961 moratorium on nuclear testing gave the world three years without atmospheric explosions and let serious and productive negotiations for a comprehensive test ban go ahead. Even though these negotiations ultimately failed, a Limited Test Ban was negotiated in 1963.

The 1972 freeze was part of SALT I and, for a limited five-year period specified in the treaty, was helpful in putting upper bounds on numbers of strategic missile launchers. It was developed in part to encourage further negotiations leading to arms reductions. The SALT II treaty was the follow-on result. In addition, even though the freeze was only for five years, the agreed launcher limitations have been held to ever since and are still being obeyed by the United States and the Soviet Union. Still another useful freeze was President Nixon's unilateral moratorium on U.S. efforts toward biological weapons, which helped obtain the international convention of 1972 prohibiting development of these weapons.

# ADVICE
# AND
# DISSENT

# Science and Morality

## WITH DONALD McDONALD

A n interview conducted by Donald McDonald in connection with a study of the American character by the Center for the Study of Democratic Institutions.

*Dr. Bethe, what is the general condition of science and the scientific community in this country as compared to what it was when you came to the United States 27 years ago? Have you seen any significant changes in the scientific environment?*

BETHE: In some respects, conditions have become better. When I arrived at Cornell, it was difficult to get enough money to run the physics department. It was very difficult for us to construct our first cyclotron, which was one of the first in the country. It cost, I think, $3,000. Today $3,000 is pin money. We use it in this laboratory in a day. So, with regard to money, things have become much easier, and this is important for the pursuit of our work.

But there are other things that go along with it. One of these is the distraction. For instance, I spent four days last week in Washington. This is a little more than usual, but I would say I spend about half of my time on things that do not have any direct connection with my science. It is somewhat worse for older people like me, but even the young people feel the external pressures to do things other than their scientific work. This clearly is not helpful.

*What are some of these pressures?*

BETHE: Let me talk about myself, that is always easiest. I have a lot to do with government committees. I used to be a member of the President's Science Advisory Committee. That committee has been assigned and has also made for itself great responsibility, particularly in the field of military technology. I sit on a large number of its subcommittees, and they take a great deal of time. Nowadays I join a committee only if I have the feeling that something good comes from it, that it has some influence on the decisions that are actually made. This particular committee is influential. There are many other committees that have little influence. Many of my colleagues are involved with such committees, as I was in the past.

*Does the administration of the tremendously increased funds for scientific research in our universities create a great demand on the time of the scientists these days?*

BETHE: Very much so. Not so much on my time because I have nothing to do with the distribution of research funds. But many of my colleagues sit on committees that must advise, say, the National Science Foundation.

*In this greatly increased financial support of scientists and scientific work accompanied by a corresponding increase in public understanding of science or its importance?*

BETHE: This varies greatly. In the executive branch of the government, there is now, I think, a great deal of understanding. I think the president and the secretary of defense recognize very strongly the value and the limitations of science. It is important that the limitations are understood too. Anybody who thinks that science can just take any order and fill it, obviously doesn't understand science. There is some vagueness and ignorance about science in some members of the Congress and in some committees of Congress. This is very bad. It is also present in some lower echelons of the Defense Department. Many Air Force officers, for instance, believe that science can do absolutely anything, that it can provide an offensive power that can break all resistance and a defensive power that can defend against all attack.

*Is there a paradox here? In what sense is science an integral part of our culture if the public still does not know how to participate in it? Are you saying that the works of science permeate our culture but that our understanding of it is faulty?*

BETHE: That is the idea. I think you have said it better than I could have said it. The real work of science goes through all our culture. And, of course, in terms of technology we see it all around us; it is the dominant feature in our life. Yet people don't try to understand the basic ideas of science.

*If there is a problem of communication between the scientists and the general public, is part of this problem a fear about the dehumanized and depersonalized work of scientists? In its most extreme form the work of the Nazi scientists, the experiments conducted on human beings, would be an example of what strikes fear in the hearts of many people. Is there, in fact, an unconcern among many scientists for human values? Do they respect only the clinical?*

BETHE: I don't want to talk about the medical and biological experiments the Nazis indulged in. They were horrible beyond belief. I suppose they were an expression of the completely inhuman attitude of the Nazi regime.

*That was more Nazi than scientific, then?*

BETHE: I think so. I think we should talk about the present American scene. Again, I would like to talk about the physical sciences because this is what I know. Most of the physicists who have had contact with atomic weapons are deeply disturbed. Most of them are very sensitive to moral values. At the same time, it is difficult for them to know what to do. There is enormous public pressure on scientists to do in the laboratories the utmost that science is capable of doing.

*There is no uniform response of American scientists to this problem?*

BETHE: The responses of the nuclear physicists range all over the map. Some of them say they don't want to have anything to do with atomic weapons. They want to work on pure science and they wish everybody would do the same. The Society for the Social Responsibility of Scientists, which is mostly a Quaker organization, recommends essentially that view. At the other end, there are people who do exactly what the government or the Air Force or the Atomic Energy Commission tells them to do. They try to invent the deadliest weapons possible and avoid thinking about the consequences. There are people who go beyond even this and some of them have great influence. They fan the flames and say we must do far more than we have done, that we must have more people working to improve atomic weapons and must spend more money on this, that the country must be geared totally to this one thing. They tell the American people of possible dangers and of weapons urgently needed to combat them, but they never estimate the probability that such dangers will occur nor do they show that the proposed weapons would be useful or even realizable.

*Is there any way out of this? Does it all come down to the individual scientist and the kind of response each scientist in his own conscience feels he must make?*

BETHE: I think what has developed is a sort of democratic process within the scientific community. There are people of all persuasions in that community. Scientists can exert some influence on the public by speeches, articles, letters to journals. Interacting with the scientists is the military community—the Air Force, Army, Navy, and Atomic Energy Commission and the rest of government, including congressional committees.

Together, these committees and the scientists have some influence on the public through the various hearings published in the newspapers. It is a complicated process, amorphous and very hard to analyze. In the nature of things, the congressional committees involved with scientists are mainly those whose task it is to foster weapons development. So, scientists who advocate development of weapons without restraint find a very ready public, while those who warn against the dangers of an unlimited arms race have no similarly ready-made audience or advocates, but are received with hostility by many members of the Washington community.

*On all levels of the Washington community?*

BETHE: No. This does not go for the leaders of government either under Eisenhower administration or the present administration. Things are all right generally at the high level, but they are very far from all right at the lower levels, and it is the lower levels generally that influence the public through the press. You hear in the newspapers most from those scientists who are strongly in favor of further military technological development and who, in their endeavor to please, sometimes talk about things that really do not exist and are years in the future, perhaps even impossible, as though they were realities or at least imminent.

*They are given to extravagant statements in the interest of their own rhetoric?*

BETHE: Not so much in the interest of their own rhetoric as in the interest of achieving unlimited weapons development, especially in their own special field. In my opinion, weapons development should not be unlimited, but in each particular case the government should examine whether the proposed weapon fits into our general military plans, whether it will tend to stabilize the world situation, etc. While some of my colleagues believe that every technological development that is feasible should actually be carried out, I believe that society—that is, in military matters the government—should make a conscious choice in each case whether the development is desirable, and whether to go ahead with it.

*The public is naturally confused when a letter appears in* The New York Times *from Edward Teller saying nuclear tests should be resumed in the atmosphere and then a few days later another letter appears from Linus Pauling stating exactly the opposite. If the scientists themselves differ so radically—and I realize that scientists cannot be completely certain on all things—how are the people supposed to be able to arrive at sound public opinion on these issues?*

BETHE: They can't; it is impossible. How can a government decide? This is one of the greatest problems of our age. The difficulty is not that the technological answers are hard to understand; they are terribly easy to understand. But many cannot be published for reasons of security. We have an elaborate system of secrecy, and rightly so. There are certain facts about our missile forces and our atomic weapons that we simply cannot state. I can understand completely the confusion of the public when they hear Teller say we have only begun to develop nuclear weapons and they hear me say we are essentially at the end of that development and anything that comes from now on is not very important. How can they tell? It is impossible.

*I suppose the best that can be done under the circumstances is for the public to have confidence in the administration in Washington, provided the administration consults and listens to all the experts it should be consulting with. I think you have said that scientists, among others, should be consulted before grave government decisions are made. Is the relationship between the science advisor to the president a real relationship with the president rather than a paper relationship?*

BETHE: It is a real one. I think in every case the advisor has seen the president at least once a week. I know that Jerome Wiesner sees President Kennedy oftener than that, or talks to him on the phone. It is a close relationship, not as close as some of the political relationships of the president, to be sure, but certainly close enough for the president to have scientific advice right next to him. And he does make use of it.

*Does the committee get full cooperation from other members of the government?*

BETHE: Yes, and this is important. The committee has been able to get information and to form its own opinions on every question it considers of any importance. Any question the committee wishes — whether of military or of general scientific interest — it may consider. It has access to all data and opinions in various branches of government. It has subcommittees on every important problem, from missile development to disarmament, and its recommendations generally are listened to.

Clearly, the main interest of the government in technology, for better or worse, is in the military area. In the subcommittees of the Science Advisory Committee an effort is always made to bring in scientists of different opinions. In these subcommittees, agreement is usually reached on scientific problems of development, although members may disagree strongly on political questions. I think this is a good way to proceed, and it is interesting that the technical situation is usually clear enough that agreement is reached in spite of the different political inclination of members. However, it is far from satisfactory for the public because it goes on behind closed doors.

*Do you think that scientists should be asked or allowed to express moral or social or political viewpoints in these advisory sessions with government leaders?*

BETHE: It is unavoidable in the advisory sessions.

*To what extent are scientists qualified to give such advice? To what extent are scientists able to speak out with some maturity on the philosophical, social and political problems of our time?*

BETHE: This is a very difficult question. Let me start from the other end. Only very few scientists are involved in actually advising the president or the secretary of defense, or whoever it might be. Usually these are scientists in whom the political official has confidence. They are also usually the ones who have some general thoughts about the world. However, in these advisory sessions, the scientist is mainly called upon to express his opinions on technical matters, as distinguished from the session we have among ourselves where there is much talk about general problems.

I think most intelligent scientists get their human values and their humanistic education more or less on their own. I don't believe you need much formal training for it. Some of the scientists are deeply interested, some are not. Those who are interested will broaden their competence. They talk to people in other disciplines, they read, they think. Being a scientist does not by itself disqualify a person for maturity in other aspects of life. Some scientists are primarily interested in their scientific work. I think that is their privilege and they make a great contribution. Presumably they will not become the science advisor to the president. Still, they may be very useful on technical subcommittees.

*Is there any problem of communication between scientists on the one hand and other professionals in the university, or in society at large, on the other hand? Is there much of a continuing dialogue between scientists and philosophers, or between scientists and political scientists?*

BETHE: Many scientists are extremely interested in politics and they talk to political scientists. On the whole, there is not much difficulty here. I have found great understanding among political scientists of the technological side of life. It is easy to agree with them on what kind of world we would all like to live in and how we would like to change the present world.

*I wonder if we could return for a moment to the moral problem, moral dilemma really, of the scientist caught up in the development of weapons technology. You have said that the scientists are now listened to more atten-*

*tively in the government than they used to be before Sputnik. But this does not resolve the moral dilemma for the scientist, does it? Are there any other things the scientist can do to resolve his moral problem with regard to the development and use of nuclear weapons?*

BETHE: I would hate to see this become too channelized and regularized. I think it is good that every scientist reacts more or less in his own way. One major problem which one faces as a scientist lies in the difference in approach to the problem-solving process between scientists and many non-scientists. For instance, when one testifies before a congressional committee, one often has the impression that the purpose of the hearing is not to search out the facts and then reason a solution, but that the solution has been determined and the hearing will now put such facts on the record as will support the solution. One might say they are not gathering facts but arguments for their position.

*All you are asking is that Congressmen and other government people be just as open to reality as scientists?*

BETHE: Yes. I think it would be good for appropriate congressional committees to have a scientist on their staff so that they can call on him in privacy for explanations in depth. I also think that congressional committees should become responsive to the whole range of scientific testimony, that they listen carefully with open minds.

*You have said that scientists who have worked on the hydrogen bomb know much more vividly the destructiveness of such a weapon than the laymen. Is this one of the special contributions that the scientist can make to the common good—that, since he has looked into the hell of the hydrogen bomb, he can advise the government with a realism others lack?*

BETHE: The scientist looked into the hell of the bomb long before anybody else did. One of the things that troubles me is that nobody believes us when we predict the hell, and that even the responses we scientists make to the hell, both inside and outside government, are not appropriate to the magnitude of that hell. The response the government made in 1949 and 1950 to go ahead and develop the bomb was natural and perhaps even correct. It was said then that we were in a cold war and we had to develop the hydrogen bomb because the Russians would develop it. Well, they sure did. But there would be no security in such a race. And I think it is obvious now that weapons are completely out of proportion, that they no longer have any function as a combination of foreign policy.

*Is the moral here that the president, who must make the final decision in the development and use of these weapons, will have a better chance of decid-*

*ing correctly if one of the components he must weight is the realistic under-standing of the nature of the weapons, an understanding that perhaps only scientist can convey to him?*

BETHE: Very definitely. It is sometimes very difficult to do this, not with regard to the president but more with regard to the lower echelons in government. However, gradually an understanding of the implications of the hydrogen bomb is spreading and statements by scientists are being understood. They are understood better because the Russians have exploded a 60-megaton bomb. Now, at last, the people are frightened. I would have liked them to be frightened to such an extent that we would have taken some other action. I still think that the proper action would have been to talk to the Russians and try to conclude an agreement not to develop such a weapon.

*With appropriate safeguards and inspection for self-protection?*

BETHE: Yes. For this, you would not have needed inspection because, while you would not have been able to tell whether the Russians were honoring an agreement not to work on the H-bomb, you certainly could have told whether they had exploded or not.

*Did we have enough deterrent capacity in 1950 that we would still have been protected had the Russians made a hydrogen bomb?*

BETHE: Yes. Against the first hydrogen bomb we had plenty of deterrent capacity. But then, of course, the question arises, what would we have done had they tested a hydrogen bomb after all. What do you do when such an agreement is broken? You still don't want to go to war. All you can do is follow suit and escalate your weapons, which doesn't do anybody any good. So far, we have not found a solution to this dilemma. Perhaps there is no satisfactory solution.

*Is there a paradox here? We have all this power, but we are also frightened to death of the consequences or of the unfathomable future. Has technology so changed our life that we have become, for example, a nation of vehicle-transported people who can no longer walk a couple of miles without becoming exhausted? Are we relying more and more on the machine in our life and less and less on our own human resources?*

BETHE: I believe so. And I think many of the things I have talked about are really technological things, rather than scientific things. Technology has had a tremendous influence on our lives, particularly war technology, I am sorry to say. I am sure war technology has greatly perturbed people. It is ever present. We have not found the right response. It cannot come from science; it must come from examining human values.

# Back to Science Advisers

## WITH JOHN BARDEEN

S cience increasingly transforms the military, political and economic landscape in which governments must operate. Nevertheless, since 1972, scientific advice to the United States government has been remarkably haphazard. For that and other reasons, the nation is embarked on vast programs based on the misconceptions that we have an unlimited supply of scientific talent and that there need be no relationship between cost and benefit.

This prodigality has a hidden price: it is destroying our ability to compete in international markets which we created. We must recapture our traditional pragmatism, or the foundations on which our security rests will crumble away.

We once had a sound scientific advisory apparatus. Established by President Dwight D. Eisenhower, it was headed by a full-time science adviser who was chairman of the President's Science Advisory Committee, composed of prominent scientists and engineers whose appointments were not correlated to Presidential elections. This system provided advice relatively uncontaminated by personal ambition and political bias.

The committee played a crucial role in many national security initiatives, including establishment of the post of director (now undersecretary) of defense for research and engineering, and also the Defense Advanced Research Program Agency. It supported the development of missile-carrying submarines, the most survivable part of our strategic forces. It fostered innovations that led to surveillance of the Soviet Union by planes and satellites, and it started the research that led to today's excellent capability to detect underground nuclear weapons tests.

In 1972, President Richard M. Nixon liquidated the entire Science Advisory Committee organization because it had opposed two of his pet projects: deployment of antiballistic missile defenses and construction of a supersonic transport. Nixon's subsequent about-face on the antiballistic missile, and the bitter English–French experience with the Concorde, soon confirmed the committee's judgments. Though the post of science adviser was eventually reestablished, it never regained the status it had when it was backed by a body with the standing of the committee.

The Strategic Defense Initiative provides a telling example of what can happen when a technology program is launched without proper technical advice, President Reagan's "Star Wars" speech was prepared without consultation with experts in the Pentagon or his own science adviser, George Keyworth, until only a few days before the speech was given. And while the White House Science Council had a panel examining related technologies, and had met five days before the speech, it was not consulted.

At the time, there were few scientists who thought there was any prospect of fulfilling the president's dream of making nuclear weapons "obsolete." Three years and billions of dollars have not changed that consensus. A recently declassified report refutes claims by members of the administration that the Initiative has scored "monumental breakthroughs." Based on interviews with many scientists at the weapons laboratories, the report concludes that their Initiative research "has resulted in a greater understanding of program difficulties, which are much more severe than previously considered." The scientists themselves resented the fact that "the progress their research has achieved has been inflated."

Perhaps the country can afford this $30 billion gamble. But top-flight scientific manpower is a rare commodity, and we cannot squander so much of it on nonproductive endeavors. Our manned space program illustrates this point. While Apollo was a great technical achievement, its cost was far greater than the dollars spent. Superior technical talent tended to gravitate to Apollo and the military programs, at the expense of civilian industries, allowing Japan to establish its dominance in consumer electronics and other markets.

Unfortunately, it has taken the Challenger disaster to bring the scientific community's concerns with our space program to the attention of the public and the government. In particular, the great majority of space missions, whether military, scientific, or commercial, would be cheaper and infinitely safer without man in space. And the tragedy itself shows what can be in store when technical judgments become subservient to schedules and politics.

It we are not to commit further follies, we shall have to recreate a scientific advisory system that has sufficient independence and prestige to give advice that is politically unpalatable. An Eisenhower advisory committee, even if composed largely of scientists from the weapons labs, would have made it perfectly clear to Reagan that his vision of defending cities against nuclear attack had no basis in scientific knowledge.

# NUCLEAR POWER

# The Necessity of Fission Power

The quadrupling of the price of oil in the fall of 1973 came as a rude, but perhaps salutary, shock to the Western world. It drew attention to the fact that oil is running out, and that mankind must turn to other fuels, to strict energy conservation, or to both.

The price increase was not entirely unjustified. From 1950 to 1973, the price of oil, measured in constant dollars, had declined steadily. Moreover, it has been estimated that if world oil production were to continue to increase at the same rate that it has in the past two decades, the upward trend could persist only until about 1995; then the supply of oil would have to drop sharply. Accordingly, the oil-producing countries must see to their own economic development while their oil lasts so that they can rely on other sources of revenue thereafter. At the same time, the rest of the world must take measures to become less dependent on oil—particularly imported oil—while there is still time.

What would it take for the U.S., which currently gets more than 15 percent of its energy in the form of imported oil, to become "energy independent?" In a report issued in June 1975, the Energy Research and Development Agency (ERDA) outlined its plans for the U.S. to achieve this goal. The ERDA projections are expressed in terms of quads, or quadrillions ($10^{15}$) of British thermal units (B.t.u.). According to ERDA, the drive to achieve energy independence calls for a two-pronged approach. First, the U.S. must be technologically geared not only to expand the production of its existing principal energy resources (oil, gas, coal and uranium), but also to develop several new energy sources. Second, a major energy-conservation effort must be initiated both to reduce total energy consumption and to shift consumption to sources other than oil. Only if both remedies are successfully applied can energy independence be achieved—and then it can be achieved only by 1995. Without any new initiatives, the need for imported oil will rise steadily from about 12 quads at present to more than 60 in the year 2000. At current prices, the importation of that much oil would cost about $120 billion, compared with $25 billion in 1974, an increase of $95 billion.

Now, $95 billion may not sound like a gigantic sum when the 1976 fiscal year's federal budget deficit is projected to be about $70 billion. The economics of international trade, however, is a different matter. Even a $10 billion trade deficit has a major effect on the stability of the currency. It is almost impossible to think of exports that could bring in an additional $95 billion. Besides, if current trends are allowed to continue, the U.S. would take about 30 percent of the world's oil production when that production is at its maximum. Clearly, it is critical that the U.S. not follow this course.

What is critical for the U.S. is a matter of survival for Japan and the countries of western Europe. After all, the U.S. does have substantial amounts of oil and gas and plenty of coal. Japan and Italy have none of those fuels. England and Norway will have a limited domestic supply of oil in a few years, but other countries of western Europe have no natural oil resources of their own and have limited amounts of coal. If the U.S. competes for scarce oil in the world market, it can only drive the price still higher and starve the economics of western Europe and Japan. The bankruptcy of those countries in turn would make it impossible for the U.S. to export to them and thus to pay for its own imports.

For the next five years or so there is only one way for the U.S. to make measurable progress toward the goal of energy independence, and that is by conserving energy. There are two kinds of energy conservation. One approach is to have the country lower its standard of living in some respects, for example by exchanging larger cars for smaller ones. This measure has been widely accepted, probably at some cost in safety. To most Americans, however, it appears undesirable to continue very far in this direction.

The other approach to conservation is to improve the efficiency with which energy is consumed. A number of useful suggestions have been made, such as insulating houses better, increasing the efficiency of space-heating and water-heating systems, improving the way steam is generated for industry, and upgrading other industrial processes. Conversions of this type require substantial investment, and their cost-effectiveness on a normal accounting scheme is not clear. Much leadership, public education and tax or other incentives will be needed to realize the potential for increased efficiency. If all these things are provided, the total energy consumption of the U.S. in the year 2000 could be reduced from 166 quads to 120.

ERDA predicts that if, at the same time, the generation of electricity from coal and nuclear fuel is allowed to expand as it is needed, the U.S. can achieve an intermediate trend in oil imports: a satisfactory decline in the first 10 years, followed by a rise until oil imports are higher in the year 2000 than they are now. Energy independence will not have been achieved by that course either.

In all three ERDA projections it is assumed that the U.S. will move gradually from liquid fuels (oil and gas) to solid fuels (coal and uranium). For example, in President Ford's State of the Union Message in January 1975, the actual contribution of various fuels to our energy budget in 1973 was presented along with the president's aims for 1985 and the expected situation in 1985 if no action is taken. The latter situation would require the importation of 36 quads of oil, in fair agreement with ERDA's prediction of 28 quads for 1985.

The Ford projection envisions a total U.S. consumption of 103 quads in 1985, 28 quads more than in 1973. Since much of the added energy would go into the generation of electricity, with a thermal efficiency of 33 to 40 percent, however, consumable energy would increase by only 17 quads, or 26 percent. Taking into account an expected 22 percent increase in the working population during that period, the consumable energy per worker would stay roughly constant.

The Ford message projects that domestic oil production will increase by seven quads by 1985 and that natural-gas production will decrease by only two quads, in spite of the fact that in the U.S. oil production has declined in the past two years and natural-gas discoveries have run at less than half of consumption for the past eight years. The ERDA report agrees that by stimulating the domestic production of oil and gas, the U.S. could attain just about the total production figure used by the president, 53 quads, with gas somewhat higher than his estimate and oil lower.

Of course, the country would be depleting its resources more rapidly and would have to pay for it by having less domestic oil and gas in the years after 1985. The proposed stimulation of domestic oil and gas production, however, would provide the breathing space needed to bring other forms of energy into play. The only energy resources the U.S. has in abundance are coal and uranium. Accordingly, President Ford calls for a massive increase in coal production, from 600 million tons in 1973 to 1,000 million tons in 1985. Meanwhile, the administration's energy program calls for the building of 200 nuclear-fission reactors with an energy output equivalent to about 10 quads.

Coal should certainly be substituted for oil and gas in utilities and in other industrial uses wherever possible. The conversion of coal into synthetic gas or oil is essential; demonstration plants for these processes and price guarantees should be given the highest priority. The same applies to oil from shale.

Coal cannot do everything, however, particularly if it is used intensively for making synthetic fuel. The U.S. needs another, preferably nonfossil, energy source. The only source that is now sufficiently developed to play any major role is nuclear fission. Thoughtful people have raised a number of objections to nuclear-fission reactors, which I shall discuss below, but first let me review some of the alternative energy sources that have been suggested.

Nuclear fusion is the energy source that has most strongly captured the imagination of scientists. It is still completely unknown, however, whether useful energy can ever be obtained from the fusion process. It is true that both stars and hydrogen bombs derive their energy from the fusion of light atomic nuclei, but can such energy be released in a controlled manner on earth? The requirements for accomplishing the task are tremendous: a mixture of heavy-hydrogen gases must be brought to a temperature of about 100 million degrees Celsius and kept there long enough for energy-releasing reactions between the hydrogen nuclei to take place at a rate sufficient to yield a net output of energy.

The most obvious way to try to satisfy this condition is by magnetic confinement. At 100 million degrees hydrogen is completely ionized, and the positively charged nuclei and negatively charged electrons can be guided by magnetic fields. Since the early 1950s, physicists in many countries have designed many intricate magnetic-field configurations, but they have not succeeded in attaining the breakeven condition. Great hopes have alternated with complete frustration. At present, the prospects seem better than ever before; a few years ago Russian experimenters developed the device named Tokamak, which has worked at least roughly according to theoretical expectations. This device has been reproduced in the U.S. with comparable success. More than $200 million has now been committed by ERDA for a much larger device of the Tokamak type, to be built at Princeton University; if that machine also fulfills theoretical expectations, we may know by the early 1980s whether or not power from fusion is feasible by the Tokamak approach.

There have been too many disappointments, however, to allow any firm predictions. Work on machines of the Tokamak type is also going forward in many other laboratories in the U.S., in the U.S.S.R. and in several countries of western Europe. If the problem can be solved, it probably will be. Money is not the limiting factor: the annual support in the U.S. is well over

$100 million, and it is increasing steadily. Progress is limited rather by the availability of highly trained workers, by the time required to build large machines and then by the time required to do significant experiments. Meanwhile, several alternative schemes for magnetic confinement are being pursued. In addition, there are the completely different approaches of laser fusion and electron-beam fusion. In my own opinion, the latter schemes are even further in the future than Tokamak.

Assume now that one of these schemes succeeds in the early 1980s. Where are we then? The problem is that the engineering of any large, complex industrial plant takes a long time, even after the principle of design is well known. Since preliminary fusion-power engineering is already under way, however, it is a reasonable hope that a prototype of a commercial fusion reactor could operate in about the year 2000, and that fusion might contribute a few percent of the country's power supply by the year 2020.

Solar power is very different. There is no doubt about its technical feasibility, but its economic feasibility is another matter. One should distinguish clearly between two uses of solar power: the heating of houses and the production of all-purpose power on a large scale.

Partial solar heating of houses may become widespread, and solar air-conditioning is also possible. ERDA is sponsoring the development of model solar-heated houses. Private estimates for solar-heating systems, for a "standard" house of 1,200 square feet, run between $5,000 and $10,000 in mass production, compared with about $1,000 for conventional heating systems. With such an installation one might expect to supply about 50 percent of the house's heating requirements (more in the South, less in the North, particularly in regions of frequent cloud cover). In any case, an auxiliary heating system supplied with gas, oil, or electricity must be provided; otherwise the cost of the solar-heating system becomes exorbitant.

ERDA estimates that 32 million new households will be established between 1972 and the year 2000, and that they will then comprise about a third of all dwelling units. If all the new units are equipped with solar heating, it would require a private investment of $150 to $300 billion. The heating requirement for all residential units in 1973 was close to 10 percent of the country's total energy consumption, and that fraction is likely to remain about the same. Some of the new dwelling units will not use solar energy, but let us assume (optimistically) that an equal number of older houses will be converted to solar heat. In that case, a third of all houses would derive on the average about half of their heat from the sun, which would then supply somewhat less than 2 percent of the country's total energy needs. This contribution would be helpful but clearly would not be decisive.

The use of solar heat on a large scale for power generation is something else again. (Here I shall assume electric power, but the situation would not be essentially different if the energy were to be stored in fuels such as hydrogen.) Of the many proposals that have been made, the most practical in my opinion is to have a large field (perhaps a mile on a side) covered by mirrors, all of which reflect sunlight to a central boiler. The mirrors would be driven by a computer-controlled mechanism; the boiler would generate electricity in the conventional manner. At least three separate groups, supported by ERDA, are working on this kind of project. The best estimates I have heard give about $2,500 per installed kilowatt (power averaged over the 24-hour day) exclusive of interest and the effects of inflation during construction. On the same basis nuclear-fission reactors cost about $500 per kilowatt, so that solar power is roughly five times as expensive as nuclear power.

That cost estimate may sound high, but a little thought will show that it is not. First of all, the sun shines for only part of the day. On a sunny winter day in the southern U.S. one square mile of focused mirrors is just about enough to generate an average of 100 megawatts of electric power at a cost of about $250 million. To achieve that output the full heat of the sun must be utilized whenever it shines. At noon such a system would generate about 400 megawatts; near sunrise and sunset it would generate correspondingly less; at night it would generate none. To get an average of 100 megawatts one must have equipment to generate 400 megawatts, so that the generating equipment (boilers, turbines, and so on) would cost roughly four times as much as they would in a comparable nuclear or fossil-fuel power plant. To this total cost must be added the cost of storing the energy that will be needed at night and on cloudy days. (The means of storage is so far a largely unsolved problem.)

Assume now that half of the cost is allotted to the mirrors and their electronic drive mechanisms; that would amount to $125 million for a plant of one square mile, or less than $5 per square foot. It is hardly conceivable that the mirrors and their drives could be built that cheaply, even in mass production, when a modest house costs $30 a square foot. I conclude therefore that all-purpose solar power is likely to remain extremely expensive.

Although it seems clear that solar power can never be practical for western Europe and Japan, the countries that need power most urgently, it might be just the right thing for certain developing countries, provided that the capital-cost problem can be solved. Many of those countries have large desert areas, rather modest total energy needs and abundant cheap manpower, which is probably required for the maintenance of any solar-power installation.

In addition to the alternative energy sources discussed above, a variety of other schemes have been suggested, such as harnessing the wind or the tides, burning garbage or agricultural wastes, converting fast-growing plants into fuels such as methane or tapping the earth's internal heat. Each of these approaches presents its own special difficulties, and at best each can make only a minor contribution toward the solution of the energy problem.

I do not mean to imply that work on alternative-energy projects is worthless. On the contrary, I believe that research and development on many of them should be pursued, and in fact ERDA is stepping up this type of work, I want to emphasize, however, that it takes a very long time from having an idea to proving its value in the laboratory, a much longer time for engineering development so that the process can be used in a large industrial plant and a still longer time before a major industry can be established. Certainly for the next 10 years and probably for the next 25 years the U.S. cannot expect any of the proposed alternative energy schemes to have much impact.

For all these reasons I believe that nuclear fission is the only major non-fossil power source the U.S. can rely on for the rest of this century and probably for some time afterward. Let us now examine the objections that have been raised against this source of power.

Some concern has been expressed over the fact that nuclear reactors in routine operation release radioactivity through outflowing liquids. According to the standards originally set by the Atomic Energy Commission (AEC) and now administered by the Nuclear Regulatory Commission, these releases must be kept "as low as practicable," and under no circumstances must the additional radiation exposure of a person living permanently near the fence of the power plant be greater than five millirem per year. Most modern fission power plants release far less than this limit. For the purposes of comparison, an average person in the U.S. receives 100 millirem per year in natural radiation (from cosmic rays, radioactivity in the earth, and in buildings and radioactive substances inside his body) and an average of about 70 millirem per year from diagnostic medical X rays. It has been estimated that in the year 2000 a person living in the U.S. would, on the average, receive an additional tenth of a millirem from nuclear reactors if 1,000 of them are deployed. Chemical plants for reprocessing the nuclear fuel may add a couple of tenths of a millirem, but the Nuclear Regulatory Commission is tightening the regulations further. In view of these very small numbers, the controversy over the routine release of radioactivity, which was strong in the 1960s, has pretty much died down.

A more popular fear at present is that a reactor accident would release catastrophic amounts of radioactivity. Here it must be said first of all that a reactor is not a bomb. In particular, light-water reactors, which make up the bulk of U.S. reactors at present, use uranium fuel with a readily fission-able uranium-235 content of only 3 percent. Such material, no matter how large the amount, can never explode under any circumstances. (For breeder reactors, which can only come into operation about 20 years from now, the argument is slightly more complicated.)

It is, however, conceivable that a reactor could lose its cooling water, melt, and release the radioactive fission products. Such an event is extremely unlikely, and one has never happened. There are at least three barriers to such a release. The radioactive fission products are enclosed in fuel pellets, and those pellets have to melt before any radioactivity is released. No such "meltdown" has occurred in nearly 2,000 reactor-years of operation involving commercial and military light-water reactors in the U.S. Moreover, even if there were to be a meltdown, the release of radioac-tivity would be retarded by the very strong reactor vessel, which typically has walls 6 to 12 inches thick. Finally, once this reactor vessel melts through, the radioactive material would still be inside the containment building, which is equipped with many devices to precipitate the volatile radioactive elements (mainly iodine, cesium and strontium) and prevent them from escaping to the outside. Only if very high pressure were to build up inside the reactor building could the building vent and release major amounts of radioactivity. The chance of that happening is extremely small, even in the event of a meltdown.

One may nonetheless ask: Exactly how likely is such a reactor accident? Obviously, it is very difficult to estimate the probability of an event that has never happened. Fortunately, most of the conceivable failures in a reactor do not lead to an accident. Reactors are designed so that in case of any sin-gle failure, even of a major part of the reactor, the reactor can still be safely shut down. Only when two or more essential elements in the reactor fail simultaneously can an accident occur. This makes a probabilistic study possible; an estimate is made of the probability of failure of one important reactor element, and it is then assumed that failures of two different ele-ments are independent, so that the probability of simultaneous failure of the two is the product of the individual probabilities. This, however, is not always true. There can be "common mode" failures, where one event trig-gers two or three failures of essential elements of the reactor; in that case, the probability is the same as that of the triggering event, and one does not get any benefit from the multiplication of small probability numbers. The probability of such common-mode failures is, of course, the most difficult to estimate.

Working on the basis of these principles, a Reactor Safety Study commissioned three years ago by the AEC estimated the probability of various types of reactor accident. The results were published in draft form in August 1974, in a document that has come to be known as the Rasmussen report, named for the chairman of the study group, Norman C. Rasmussen of the Massachusetts Institute of Technology. The final report was published October 1975.

The methods applied in the Rasmussen report have been used for several years in Britain to predict the probability of industrial accidents. Experience has shown that the predictions usually give a frequency of accidents somewhat higher than the actual frequency. Several groups, including the Environmental Protection Agency and a committee set up by the American Physical Society, have since studied various aspects of the problem and have come out with somewhat different results. Those differences have been taken into account in the final Rasmussen report; the most important of them will be discussed here.

The basic prediction of the Rasmussen report is that the probability of a major release of radioactivity is about once in 100,000 reactor-years. (Common-mode failures were found to contribute comparatively little to the total probability.) Such an accident would involve the release of about half of the volatile fission products contained in the reactor. A release of that scale would have to be preceded by a meltdown of the fuel in the reactor, an event for which the report gives a probability of once in 17,000 reactor-years. Finally, the report predicts that the water coolant from a reactor will be lost once in 2,000 reactor-years, but that in most cases a meltdown will be prevented by the emergency core-coolant system.

There is at least some check on those estimates from experience. For one thing, there has never been a loss of coolant in 300 reactor-years of commercial light-water-reactor operation. Furthermore, there has never been a fuel meltdown in nearly 2,000 reactor-years of commercial and naval light-water-reactor operation. If Rasmussen's estimate were wrong by a factor of 20 (in other words, if the probability of a meltdown were once in 850 reactor-years), at least one meltdown should have occurred by now.

What would be the consequences in the extremely improbable event of a major release of radioactivity? The immediate effects depend primarily on the population density near the reactor and on the wind direction and other features of the weather.

For a fairly serious accident (one that might take place in a million reactor-years) Rasmussen estimates less than one early fatality but 300 cases of early radiation sickness. He also predicts that there could be 170 fatalities per year from latent cancers, a death rate that might continue for 30 years, giving a total of some 5,000 cancer fatalities. In addition there might be 25 genetic changes per year; counting the propagation of such changes through later generations, the total number of genetic changes could be about 3,000.

The number of latent cancers in the final version of the Rasmussen report is about 10 times as high as it was in the original draft report; that change was largely suggested by the study of the American Physical Society, as modified by a very careful study made by the Rasmussen group. A major release of radioactivity under average weather and population conditions (probably one in 100,000 reactor-years) would cause about 1,000 latent cancers, but it would not result in any cases of early radiation sickness.

It is obvious that 5,000 cancer deaths would be a tragic toll. To put it in perspective, however, one should remember that in the U.S. there are more than 300,000 deaths every year from cancers due to other causes. A reactor accident clearly would not be the end of the world, as many opponents of nuclear power try to picture it. It is less serious than most minor wars, and these are unfortunately quite frequent. Some possible industrial accidents can be more serious, such as explosions and fires in large arrays of gasoline storage tanks or chemical explosions. The danger from dam breaks is probably even greater.

The probability of a serious reactor accident was predicted in the Rasmussen report to be once in 10,000 years when there are 100 reactors, which is about the number expected for the U.S. in the year 1980. What if the number of reactors increases to 1,000, as many people predict for the year 2000 or 2010? The answer is that reactor safety is not static but is a developing art. The U.S. is now spending about $70 million per year on improving reactor safety, and some of the best scientists in the national laboratories are engaged in the task. I feel confident that in 10 years these efforts will improve both the safety of reactors and the confidence we can have in that safety. I should think that by the year 2000 the probability of a major release of radioactivity will be at most once in 10 million reactor-years, so that even if there are 1,000 reactors by that time, the overall chance of such an accident will still be no more than once in 10,000 years.

Taking into account all types of reactor accidents, the average risk for the entire U.S. population is only two fatalities per year from latent cancer and one genetic change per year. Compared with other accident risks that our society accepts, the risk from nuclear reactors is very small.

A special feature of possible reactor accidents is that most of the cancers would appear years after the accident. The acute fatalities and illnesses would be rather few compared with the 5,000 estimated fatalities from latent cancers in the foregoing example. The problem is that many more than the 5,000 victims will think they got cancer from the radiation, and it will be essentially impossible to ascertain whether radiation was really the cause. The average probability that the exposed population will get fatal cancer from the released radioactivity is only about 1 percent, compared with the 15 percent probability that the average American will contract fatal cancer from other causes. Will the affected people in the case of a reactor accident be rational enough to appreciate this calculation? Or would an accident, if it occurs, have a psychological effect much more devastating than the real one?

The problem of nuclear energy that is considered most serious by many critics is the disposal of nuclear wastes. Will such wastes poison the atmosphere and the ground forever, as has been charged? It is true that the level of radioactivity in a standard 1,000-megawatt reactor is very high: about 10 billion curies half an hour after the reactor is shut down. The radioactivity then decays quite quickly, however, and so does the resulting heat.

When the spent nuclear fuel is unloaded from a reactor, it goes through a number of stages. First the highly radioactive material, still in its original form, is dropped into a tank of water, where it is left for a period ranging from a few months to more than a year. The water absorbs the heat from the radioactive decay and at the same time shields the surroundings from the radiation.

After the cooling period the fuel will in the future be shipped in specially protected trucks or railcars to a chemical-reprocessing plant. (No such plant is currently in operation, but a large one is being built in South Carolina and could go into operation next year.) In the chemical plant the fuel rods will be cut open (still under water) and the fuel pellets will be dissolved. The uranium and the plutonium will be separated from each other and from the radioactive fission products. The uranium and plutonium can be reused as reactor fuel and hence will be refabricated into fuel elements. The remaining fission products are the wastes.

These substances are first stored in a water solution for an additional period to allow the radioactivity to decay further. Special tanks with double walls are now being used for that purpose in order to ensure against leakage of the solution.

After five years the wastes will be converted into solids, and after another five years they will be shipped to a national repository. Three different methods have been developed for solidifying wastes; one method now operates routinely at ERDA's reactor test station in Idaho to solidify the wastes from government-owned reactors. The solid wastes can then be fused with borosilicate glass and fabricated into solid rods, perhaps 10 feet long and 1 foot in diameter. (Approximately 10 such rods will be produced by a standard 1,000-megawatt reactor in a year.) The rods are then placed in sturdy steel cylinders closed at both ends. It is difficult to see how any of the radioactive material could get out into the environment after such treatment, provided that the material is adequately cooled to prevent melting.

There are two possibilities for the national repository. One, for interim storage, would be in an above-ground desert area; the steel cylinders would be enclosed in a heavy concrete shield to protect the external world from the radiation. Cooling would be provided by air entering at the bottom of the concrete shield, rising through the space between the steel and the concrete and escaping at the top after having been heated by about 20 °C. Natural air circulation would be sufficient; no fans are required. The proposal for such a national repository has been studied and approved by a committee of the National Academy of Sciences.

The area required for such an interim-storage repository is not large. A standard reactor produces about two cubic meters of solid waste a year. The National Academy of Sciences committee estimated that all the wastes produced by U.S. reactors by the year 2010 could be stored on a tract of 100 acres. The cost is estimated at $1.5 billion, a small fraction of the probable cost of the reactors.

The second possibility for the national repository is permanent storage deep underground. The preferred storage medium here is bedded salt, which presents several advantages. First, the existence of a salt bed indicates that no water has penetrated the region for a long time; otherwise the salt would have been dissolved. Water trickling through the storage site should be avoided, lest it leach the deposited wastes and bring them back up to the ground, an extremely slow process at best, but still better avoided altogether. Second, salt beds represent geologically very quiet regions; they have generally been undisturbed for many millions of years, which is good assurance that they will also remain undisturbed for as long as is required. Third, salt flows plastically under pressure, so that any cracks that may be formed by mechanical or thermal stress will automatically close again.

The first attempt by the AEC to find a storage site in a salt mine in Kansas was unfortunately undertaken in a hurry without enough research. (Drill holes in the neighborhood might have allowed water to penetrate to the salt bed and the waste.) Now ERDA is carefully examining other sites. A promising location has been found in southeastern New Mexico. There are roughly 50,000 square miles of salt beds in the U.S.; only three square miles are needed for disposal of all the projected wastes up to the year 2010.

The method of disposal is this: In a horizontal tunnel of a newly dug mine in the salt bed, holes would be drilled in the wall just big enough to accommodate one of the steel cylinders containing waste. It has been calculated that the cylinders could be inserted into the salt 10 years after the waste comes out of the reactor. The residual heat in the waste, five kilowatts from one cylinder, is then low enough for the salt not to crack. (The high heat conductivity of salt helps here.) If the calculation is confirmed by experiment in the actual mines, the wastes could go directly from the chemical-processing plant into permanent disposal and interim storage would be unnecessary. Otherwise the wastes would be placed for some years in the interim repository and then be shifted from there to permanent storage underground.

It seems to me virtually certain that a suitable permanent storage site will be found. It is regrettable that ERDA is so slow making a decision and announcing it, but after the difficulties with the Kansas site it is understandable that ERDA wants to make absolutely sure the next time.

Most of the fission products have short half-lives, from a few seconds to a few years. The longest-lived of the common products are cesium-137 and strontium-90, with half-lives of about 30 years. The problem is that the wastes also contain actinides: elements heavier than uranium. In the present chemical process .5 percent of plutonium and most of the other actinides go with the wastes. Plutonium-239 has a half-life of nearly 25,000 years, and 10 half-lives are required to cut the radioactivity by a factor of 1,000. Thus the buried wastes must be kept out of the biosphere for 250,000 years.

Scientists at the Oak Ridge National Laboratory have studied the possible natural events that might disturb radioactive-waste deposits and have found none that are likely. Similarly, it is almost impossible that man-made interference, either deliberate or inadvertent, could bring any sizable amount of radioactivity back to the surface.

The remaining worry is the possibility that the wastes could diffuse back to the surface. The rate of diffusion of solids in solids is notoriously slow, and experiments at Oak Ridge have shown that the rate holds also for the diffusion of most fission products in salt. Ultimately, this observation will have to be confirmed in the permanent storage site by implanting small quantities of fission products and observing their migration.

In the meantime, one can draw further confidence from a beautiful "experiment" conducted by the earth itself. It has been discovered that in the part of Africa now called the Gabon Republic there existed some 1.8 billion years ago a natural nuclear reactor. Metal ore in that area is extremely rich in uranium, ranging from 10 to 60 percent. Whereas the present concentration of uranium-235 in natural uranium is .72 percent, the concentration 1.8 billion years ago was about the same as it is in present-day light-water reactor (3 percent). The ore also contained about 15 percent water. Therefore conditions were similiar to those in a light-water reactor (except for the cooling mechanism). In the natural nuclear reactor plutonium-239 was formed, which subsequently decayed by emitting alpha radiation to form uranium-235. The interesting point is that the plutonium did not move as much as a millimeter during its 25,000-year lifetime. Moreover, the fission products, except the volatile ones, have stayed close to the uranium, even after nearly two billion years.

Assuming that plutonium is made in appreciable amounts, it must be kept from anyone who might put it to destructive use. Contrary to a widespread fear, however, there is little danger that plutonium could be stolen from a working nuclear reactor. The reactor fuel is extremely radioactive, and even if an unauthorized person were to succeed in unloading some fuel elements (a difficult and lengthy operation), he could not carry them away without dying in the attempt. The same is true of the used fuel cooling in storage tanks. The places from which plutonium might in principle be stolen are the chemical reprocessing plant (after the radioactive fission products have been removed), the fuel-fabrication plant or the transportation system between the plants and the reactor where the refabricated fuel elements are to be installed.

Transportation seems to be the most vulnerable link. Therefore it is probably desirable to establish the chemical plant and the fuel-fabrication plant close together, leaving only the problem of transportation from there to the reactor. Actually the problem of secure and safe transportation is essentially solved, at least in the U.S. The sophisticated safeguards now in force for nuclear weapons can be easily adapted for the transportation of nuclear materials. The protection of plants against theft is also being worked on and does not appear to present insuperable problems. For example, people leaving a plant (including employees) can be checked for possession of plutonium, even in small amounts, by means of automatic detectors, without requiring a body search. These direct measures for safeguarding plutonium are necessary and cannot be replaced by simple inventory-accounting procedures, which would be far too inaccurate. By ensuring that no plutonium (or fissionable uranium) has been diverted from U.S. plants, one can be reasonably confident that no terrorists in this country can make an atomic bomb (which, by the way, is not as easy as some books and television programs have pictured it).

It has been asserted that the proposed measures for safeguarding plutonium and similar measures for protecting nuclear power plants from sabotage will interfere with everyone's civil liberties. I do not see why this should be so. The workers in the nuclear plants, the guards, the drivers of trucks transporting nuclear material and a few others will be subject to security clearance (just as people working on nuclear weapons are now). I estimate their number at less than 20,000, or less than 1 percent of our present armed forces. The remaining 200 million Americans need suffer no abridgement of their civil liberties.

Plutonium has been called the most toxic substance known. The term toxicity can be misleading in this context, because it implies that the danger lies in some chemical action of plutonium. Experiments with animals have shown that it is the level of radioactivity of the plutonium that counts, not the quantity inhaled, as is the case with a chemical poison. Nonetheless, the radioactive hazard is indeed great once plutonium is actually absorbed in the body: .6 microgram of plutonium-239 has been established by medical authorities as the maximum permissible dose over a lifetime, and an amount approximately 500 times greater is believed to lead to lethal cancer.

Plutonium can be effectively absorbed in the body if microscopic particles of it are inhaled. About 15 percent of the particles are likely to be retained in the lung, where they may cause cancer. Fortunately there is little danger if plutonium is ingested in food or drink; in that case it passes unchanged through the digestive tract, and only about one part in 30,000 enters the bloodstream. Therefore effective plutonium poisoning of the water supply or agricultural land is virtually impossible.

Some opponents of nuclear power have maintained that because of the very low maximum permissible dose even small amounts of plutonium in the hands of terrorists could cause great damage. This point has been put in perspective by Bernard Cohen, who has investigated in theory the effect of a deliberate air dispersal of plutonium oxide over a city. He finds that on the average there would be one cancer death for every 15 grams of plutonium dispersed, because only a small fraction of the oxide would find its way into people's lungs. Other, soluble compounds of plutonium would be even less effective than an insoluble oxide. A terrorist who manages to steal six kilograms or more of plutonium could probably do more damage by fashioning a crude bomb from it than by dispersing it in the air of a city.

Will the spread of nuclear reactors encourage the proliferation of nuclear weapons? That in my opinion is the only really serious objection to nuclear power. The availability of fissionable material is obviously a prerequisite for making nuclear weapons. Even after the material is available, however, the manufacture of a nuclear bomb is still a massive undertaking: in each of the six countries that have so far conducted nuclear explosions, thousands of scientists and technicians have worked on the development of the weapon. Nonetheless, a number of additional countries would be capable of this effort if they wanted to make it, and if they had the material.

Many countries in need of nuclear power will soon be in the market for the purchase of nuclear power plants from any country willing to sell them. Nuclear power plants sold in international trade are usually put under the inspection system of the International Atomic Energy Agency (IAEA) in order to ensure that no fissionable material is diverted for military purposes. The IAEA needs strengthening and more money for its force of inspectors. An important additional safeguard would be to prevent the proliferation of nuclear chemical-processing plants, since it is from those plants rather than from the reactors that fissionable material could be diverted. A good proposal is that the chemical processing be centralized in plants for an entire region rather than dispersed among plants for each nation. Another approach would be to have the country supplying the reactor lease the fuel to the customer country with the requirement that the used fuel be returned.

The original fuel for a light-water reactor is mostly uranium-238 enriched with about 3 percent of readily fissionable uranium-235. If an explosive were to be made from this fuel, the two isotopes would have to be separated, a procedure that requires a high level of technology. The used fuel contains in addition some plutonium, which can be separated from the uranium by chemical procedure, a less difficult task. The resulting plutonium has a high concentration of plutonium-240 (with respect to plutonium-239), which could be used to make rather crude bombs by a country just beginning in nuclear-weapons technology. Breeder reactors contain more plutonium per unit of power, with a smaller percentage of plutonium-240. I personally would therefore recommend that breeder reactors not be sold in international trade.

Proliferation would not be prevented if the U.S. were to stop building nuclear reactors for domestic use or if it were to stop selling them abroad. Western Europe and Japan not only need nuclear power even more than the U.S. does but also have the technology to acquire it. Moreover, they need foreign currency to pay for their oil imports and so they will want to sell their reactors abroad. The participation of the U.S. in the reactor trade may enable us to set standards on safeguards, such as frequent IAEA inspection, that would be more difficult if we left the trade entirely to others.

It has been alleged that nuclear power is unreliable. The best measure of reliability is the percentage of the time a plant is available for power production when the power is demanded. This "availability factor" is regularly reported for nuclear plants and runs on the average about 70 percent. There are fewer good data on the availability of large coal-fired plants, but where the numbers exist they are about the same as those for nuclear plants.

The "capacity factor" is the ratio if the amount of power actually produced to the amount that could have been produced if the plant had run constantly at full power. That percentage is usually lower than the availability factor for two reasons: (1) some nuclear power plants are required for reasons of safety to operate below their full capacity, and (2) demand fluctuates during each 24-hour period. The second factor is mitigated by the operation of nuclear reactors as base-load plants, that is, plants that are called on to operate as much of the time as possible, because the investment cost is high and the fuel cost is low. A reasonable average capacity factor for nuclear power plants is 60 percent. One utility has estimated that at a capacity factor of 40 percent nuclear and coal-fired plants generating the same amount of electricity would cost about the same; operation at 60 percent therefore gives the nuclear plant a substantial edge.

But are not nuclear power plants expensive to build? An examination of the construction cost of such plants planned between 1967 and 1974, and expected to become operational between 1972 and 1983, shows that the cost of a 1,000-megawatt power plant of the light-water-reactor type has risen from $135 million to $730 million in this period. Closer inspection reveals, however, that a large fraction of the cost increase is due to inflation and to a rise in interest rates during construction; without those factors the 1974 cost is $385 per kilowatt of generating capacity. This figure represents a cost increase of about 300 percent, which is more than the general inflation from 1967 to 1974. The main cause must be looked for in the steep rise of certain construction costs, particularly labor costs, which rose about 15 percent per year, or 270 percent in seven years.

The cost of building coal-fired plants has risen at a comparable rate. A major factor here has been the requirement of "scrubbers" to remove most of the sulfur oxides that normally result from the burning of coal. Coal plants equipped with scrubbers may still be about 15 percent cheaper to build than nuclear plants. Any massive increase in coal production would, however, call for substantial investment not only in the opening and equipping of new mines but also in the provision of additional railroad cars and possibly tracks, particularly in the case of Western mines. If this "hidden" investment is included, the capital cost of coal-burning power plants is not very different from that for nuclear plants. Even disregarding this factor, the overall cost of generating electricity from nuclear fuel is already much less than it is for generating electricity from fossil fuel, and recent studies indicate that nuclear power will continue to be cheaper by a wide margin.

There is some truth in the charge that "nuclear power does not pay its own way," since the government has spent several billion dollars on research on nuclear power and several more billions will undoubtedly have to be spent in the future. On the other hand, the government is also spending about $1 billion a year as compensation to coal miners who have contracted black-lung disease.

It has also been said that uranium will run out soon. It is true that the proved reserves of high-grade uranium ore are not very large, and the existing light-water reactors do require a lot of uranium. If all reactors were of this type, and if the U.S. were to set aside all the uranium needed for 40 years of reactor operation, then the total uranium-ore resources of the U.S. would only be enough to start up 600 reactors, a number that might be reached by the year 2000. Beyond that date it will be important to install reactors that consume uranium more efficiently. The most satisfactory alternative to emerge so far is the breeder reactor, which may be ready for industrial operation by 1990. The breeder, in effect, extracts the energy not only from the rare isotope uranium-235 but also from other isotopes of uranium, thereby increasing the supply of uranium about sixtyfold. Even more important, with the breeder the mining of low-grade uranium ore can be justified both economically and environmentally. With these added resources there is enough uranium in the U.S. to supply 1,000 reactors for 40,000 years.

As interim alternatives, two other types of reactor are attractive: the high-temperature, gas-cooled, graphite-moderated reactor and the Canadian natural-uranium reactor ("Candu"), which is moderated and cooled by heavy water. The Candu reactor can be modified to convert thorium by neutron capture into the fissionable isotope uranium-233.


In weighing the overall health hazard presented by nuclear reactors it is appropriate to compare nuclear plants with coal-burning power plants. Recent findings indicate that even if scrubbers or some other technology could reduce the estimated health effects from coal burning by a factor of 10 (which hardly seems attainable at present), the hazard from coal would still exceed that from nuclear fuel by an order of magnitude. This comparison is not meant as an argument against coal. The U.S. clearly needs to burn more coal in its power plants, and even with coal the hazard is not great. The comparison does point up, however, the relative safety of nuclear reactors.

In sum, nuclear power does involve certain risks, notably the risk of a reactor accident and the risk of facilitating the proliferation of nuclear weapons. Over the latter problem, the U.S. has only limited control. The remaining risks of nuclear power are statistically small compared with other risks that our society accepts. It is important not to consider nuclear power in isolation. Objections can be raised to any attainable source of power. This country needs power to keep its economy going. Too little power means unemployment and recession, if not worse.

# Debate: Nuclear Safety

## RESPONSE BY FRANK VON HIPPEL

The American Physical Society's Study Group on Light Water Reactor Safety study found that the possible consequences of a severe reactor accident are substantially greater than had been concluded in the Rasmussen report. However, the basic conclusions of the Rasmussen report remain unchanged: The actual risk from reactor accidents is very small indeed.

The APS study did not investigate the probability of reactor accidents nor did it investigate the number of immediate casualties from a hypothetical reactor accident. According to the Rasmussen report, in an average situation of population density and weather, the "reference accident" — called category PWR-2 — will lead to about 300 casualties from radiation, both immediate and delayed. The APS study differs from this estimate of the possible consequences of a severe accident by taking into account certain delayed cancers. They are of two kinds:

- Delayed cancers resulting from high-level exposure of persons *close* to the reactor accident. These are mostly lung and thyroid cancers. The APS study calculated 1,100 to 5,600 cancer deaths from high-level exposure. The uncertainty is very large, the geometric mean is 2,500. This number depends on the particular location of the reactor and the weather conditions; and it is eight times the number given in the Rasmussen report.

- Delayed cancers from low-level radiation which occur *at distances between 20 and 500* miles at times later than one day after the accident. These effects are mainly due to cesium-137 which has a half-life of about 30 years and emits gamma rays. The number estimated in the APS study is 10,000, assuming a population density of 300 persons per square mile. The Rasmussen report states that the correct population density to use is not more than 165 per square mile, and perhaps even only half of this number because in many cases points 500 miles from the reactor are in the ocean. The APS number should therefore be multiplied by .55, giving 5,500 delayed deaths from cancer.

I consider this number an upper limit. First, because it should be easy to remove the cesium, at least from dwellings, by decontamination. Cesium is about the most reactive element, and forms soluble salts which probably will be washed away by rain even if no active decontamination is undertaken. Second, because most of the biomedical evidence indicates that the effect of very low-level radiation in producing cancer is much less than proportional to the total dose received. It is true that the BEIR report* recommends using the linear assumption (effect proportional to total dose regardless of dose rate) but the direct evidence, both from animal experiments and from the Hiroshima-Nagasaki experience, strongly suggests that low levels are much less effective.

Still, taking the calculated number 5,500 we get for the total number of delayed cancer deaths in the "reference accident" 6,500 to 10,500.

The probability of the "reference accident" is estimated in the Rasmussen report to be one in 200,000 per reactor year. Assuming 100 reactors of 1,000 megawatts electrical, there is then an average risk of three to five fatalities per year for the whole of the United States.

It so happens that the probability of the "reference accident" (category PWR-2) has been greatly decreased by a design change recommended as a consequence of the Rasmussen report: The main accident cause in category PWR-2 came from the presumed failure of a check valve. The manufacturer has cured this trouble, and thereby the probability of this accident category has been reduced to about 20 percent of its former value. However, if we include other accident categories and weight them with the amount of radioactivity released in each case, we get for the weighted probability of *all* accidents involving substantial release, again, the figure of one in 200,000. The estimate of three to five fatalities per year can, therefore, be considered as the total risk from nuclear accidents.

When extrapolating to the future, it should be kept in mind that reactor safety is not a static matter. Safety research will be conducted, in fact with a considerably increased budget, and there is every reason to expect that the safety will be greatly improved over the Rasmussen assumptions by the

---

*The Committee on the Biological Effects of Ionizing Radiation (the BEIR Committee) is an advisory committee of the National Academy of Sciences-National Research Council. Its 1972 report on "The Effects on Populations of Exposure to Low Levels of Ionizing Radiation" is commonly referred to as the BEIR report.

time 1,000 nuclear reactors are working. The risk at that time might be even lower than three to five fatalities per year.

Another way of looking at the APS results is given in the APS study itself: The population that is exposed to fallout from the reactor has an average probability of contracting fatal cancer due to the accident of only .1 percent. In the United States, this should be compared with a 20 percent probability of contracting cancer from other causes.

Considering all this, I conclude that the APS study has not fundamentally changed the situation. While it is true that a single large accident may lead to several thousand fatalities, most of them delayed and spread over many years, the average risk per year remains at an extremely low level.

---

## FRANK VON HIPPEL RESPONDS

One of the major findings of the American Physical Society's reactor safety study was that a nuclear reactor core meltdown accident followed by a major release of gaseous and volatile fission products to the atmosphere (that is, a containment failure) could well have exceedingly severe consequences. In the past this would not have been surprising: that is, before the Atomic Energy Commission's Reactor Safety Study (the Rasmussen report), it was assumed that this would be the case.

The Rasmussen report challenged that idea—that a nuclear reactor core meltdown accident with containment failure could have such severe consequences. The APS study group reviewed the calculations made in that report and found, using essentially the same assumptions, that two omissions in the calculations made in the AEC study had resulted in an underestimate of the average number of deaths from their "reference" accident by one or two orders of magnitude. Other major underestimates were also identified in the Rasmussen report's calculations of other categories of consequences, including genetic defects, nonfatal cancers, and groundwater contamination.

Thus, it appears that the APS study may have the dubious honor of restoring the status quo: a reactor meltdown accident with containment failure could indeed be very serious. Whereas the AEC's Rasmussen report may have unintentionally conveyed the message that we can relax about nuclear reactor safety, I believe the correct message should be that the industry and its watchdogs have to work as hard as ever at making nuclear reactors safe, and that major improvements are called for in several important areas.

While one can and should be critical of some of the calculations of the Rasmussen report, I believe that it has also done a great service by making a first attempt to put the reactor-safety debate into perspective. In fact, I would like to offer here a perspective which was inspired in part by the approach taken in the Rasmussen report.

The first basic premise is that our society obtains great benefits from electric energy. The second is that none of our current sources of electric energy are risk free. Therefore, the risks associated with nuclear-electric energy must be weighed against those of the alternatives.

In the past, the emphasis in the environmental debate relating to electrical energy focused on the routine impacts of the coal and nuclear industries. This is still true in the case of coal where the emphasis is on bad strip mining practices, the occupational hazards of underground coal mining, and the health consequences of the sulfur dioxide emissions from coal-burning plants.

In the case of nuclear power, the emphasis was on the small amount of radioactivity released routinely from nuclear power plants. The limits on the allowable releases of radioactivity from nuclear power plants have now been reduced to the point where few critics think that the issue is worth their time any more. The focus of the debate over nuclear reactors, therefore, has shifted to low-probability high-consequence events. In particular, the debate has focused on nuclear reactor core meltdown events—either accidental or as a result of sabotage.

The general impression seems to be that such potentially catastrophic events have no counterpart in our coal or oil fired energy generating technology. I think they do. Let me focus specifically on two: war and climatic change.

The United States has been involved in three major wars in the past three decades. The consequences of each of these wars was much worse than any nuclear reactor accident could be. Today one of the major sources of world tension stems from the international oil economy. An important fraction of this oil is used for electrical power generation. It might be fair, therefore, to attribute some significant share of the probability of the next major war to the use of oil for the generation of electricity.

In the case of coal, we have an ample domestic resource base, but we also have the potential to cause substantial changes in the earth's climate. The burning of only a few percent of the world's estimated resources of recoverable fossil fuels has already resulted in an increase in the amount of carbon dioxide in the atmosphere by about 10 percent. By the year 2000, it is estimated that we will have increased the carbon dioxide level in the atmosphere by another 10 to 20 percent. What will the impact be on the world's climate? Climatologists can't tell us because of large feedback effects and natural variations which are not well understood. They do tell us, however, that changes with a significant adverse impact on world agriculture *could* occur.*

The continued use of oil and coal therefore also involves—with unknown probability—events with potentially very large consequences. In fact, I think that even a major nuclear accident pales somewhat in comparison with the nightmares of war and famine which I have just conjured up.

The other comparison which may help put the nuclear accident in perspective concerns the cancers and genetic defects which would result from high or low doses of radiation. This is an area where the APS study found potentially much larger consequences than the AEC's study. It should be pointed out, however, that the number of additional casualties would be diluted in a very large population (spread over thousands of square miles downwind) and would also be spread out in time over a period of decades after the accident. As a result, the added risk to most of the *individuals* in the population downwind would be quite low—less than 1 percent, *except* in the case of thyroid damage. An increase in the cancer death rate of a few tenths of a percent and an increase in the incidence of genetic defects by a few hundredths of a percent would hardly be detectable relative to the current rates of occurrence of about 20 percent and 2 percent, respectively.

This would be a rather Strangelovian statement if it weren't for the fact that the current incidence of cancer and genetic defects may well be, in large part, man-caused. These categories of consequences of a nuclear accident would therefore not be unique at all, but would be added to a much larger pool of cancers and genetic defects which we are already inflicting upon ourselves with chemicals, medical X rays, etc.

These arguments are not intended to imply that we should drop the nuclear reactor safety issue. I offer them simply to explain why it is not so obvioius to me at this point that we should abandon nuclear energy more rapidly than fossil power—at least on the grounds of nuclear reactor safety.

In this context then, how might reactor safety be improved? In fact there seem to be some important possibilities. Consider first two of the suggestions made by the APS study: (1) improved containment design and (2) mitigation of accident consequences.

*See, for example, J. Murray Mitchell, Jr., "A Reassessment of Atmospheric Pollution as a Cause of Long-Term Challenges of Global Temperature," in *The Changing Global Environment*, edited by S. Fred Singer (Reidel, 1975).

Can the reactor containment building be designed to contain, in fact, a nuclear reactor core meltdown accident? On the one hand, there are obvious difficulties. The enormous amount of radioactive energy released by the core after shutdown threatens the integrity of the building in many ways: steam explosions and overpressure, hydrogen and carbon dioxide gas generation and, ultimately, melt-through. On the other hand, the AEC never asked the nuclear power plant designers to design a containment building with the possibility of a core meltdown accident in mind. It would appear to be desirable that this policy be changed.

There are some real opportunities for improvement. In particular, it appears that a containment building could be designed so that the important radioactive isotopes could be trapped in some sort of filter system, even if containment failed because of overpressure. A serious design effort might also produce a "core-catcher" to prevent melt-through by the core and the associated potential for massive water contamination. And, finally, underground siting would appear to offer protection to the containment building against failure, either as a result of a steam explosion inside or as a result of the impact of missiles from the outside—accidental or deliberate.

Having an improved containment structure would probably do far more for reactor safety than adding another safety system onto the reactor itself. As the findings of the Rasmussen report dramatized, new valves and plumbing introduce new accident scenarios as well as preventing the development of old ones.

Being in a position to mitigate the consequences of an accident, should one occur, would obviously also be desirable. Currently, the only mitigation strategy is evacuation—and there is a real question whether this could in fact be effective. It might not be possible to evacuate nearby people fast enough; and it would probably not be practical to evacuate the enormous number of people further away who would be exposed individually to relatively low doses.

Two of the mitigating strategies which the APS study suggested were (1) the distribution of iodide pills to block the thyroid's uptake of radioiodides, and (2) the development of practicable decontamination techniques to reduce the long-term population doses from the 30-year half-life isotope, cesium-137.

These two efforts might reduce by as much as an order of magnitude the human costs of a major release of radioactivity to the atmosphere. Once again, they could be made much more convincing than improved plumbing for the reactor.

Let us turn now, however, to the issue of improved reactor design. No group has seriously compared the safety advantages of different reactor designs. It is true that the Rasmussen report did compare the safety of the boiling water and pressurized water reactors and found that these two reactor types were almost exactly equal in safety—a remarkably improbable result incidentally! But no one has done a serious comparison of these reactors with, for example, the much different Canadian heavy-water reactor or the high-temperature gas-cooled reactor.

Such a comparison might teach us something. Comparisons are made every day of the *economic* advantages of different reactor designs; but there seems to be a gentlemen's agreement not to debate the *safety* advantages. If it ever becomes possible to discuss reactor safety rationally, an open discussion of the relative safety advantages of various design features might lead to some important design improvements.

Another concern which must be addressed is whether, as the nuclear industry grows and as the mystique of nuclear power wears off, the industry will become sloppy. The pioneers who developed this technology had real respect for its dangers. Today, there are many signs that not all those who have inherited the technology have that same respect. This means that reactor technology and regulation will have to be made more protective against fools.

Perhaps, even more importantly, nuclear technology has to be designed better against deliberately destructive acts. When today's technology was being developed, 10 to 15 years ago, no one thought about terrorists. Now, unfortunately, we have to—and we must, as a result, push our nuclear technology closer to the goal of being intrinsically safe.

The fact that none of the above suggestions were considered seriously by the AEC in the past dramatizes for me the importance of having outside watchdog groups. In fact, it might be appropriate at this point to acknowledge the importance of the efforts of the Union of Concerned Scientists— Henry Kendall in particular—in opening up the whole issue of reactor safety and in attracting the interest of those of us who organized the APS study. I have become convinced by my experiences in this area that the watchdog role must be a continuing one. As a society we cannot afford to have the nuclear energy regulators become as fat and sleepy as regulators have in other areas.

The necessary improvements in nuclear reactor safety will not be possible if we go ahead with a crash program to build all of the 1,000 large reactors and associated plants which the government envisions for the next 25 years. Similar concerns can be raised about the consequences of the projected drastic growth in our use of coal. In fact, I believe that every time we double our rate of energy consumption, we decrease our future options dramatically and increase the probability of triggering some major social or environmental instability. One might well ponder whether a society that has been unable to implement even the simplest strategies for increasing the efficiency of its energy use can really deal successfully with the more demanding challenges of modern energy technology.

A perspective on nuclear reactor safety would be incomplete if it ignored the safety issues which relate to the rest of the nuclear fuel cycle.

It is at the reprocessing plant, for example, where the zirconium sheath around the fuel will be chopped up and the ceramic fuel pellets will be dissolved, freeing enormous amounts of radioactivity in gaseous and liquid form. This is also where plutonium first becomes available for diversion or theft. When we have a better perspective on nuclear power, I believe that we may realize that it is at the reprocessing plant and not at the reactor where the key issues of safety and environmental compatibility of nuclear energy will be found. For this reason I am proposing that these issues be examined in the next outside technical review of fission power.

In summary,

- The consequences of a nuclear reactor accident or sabotage could be much more serious than indicated by the Rasmussen report.
- It does not necessarily follow, however, that the social risks associated with the nuclear reactor accidents are more severe than those associated with fossil fuels.
- But this comparison provides scant comfort and since there are major opportunities for improving reactor safety, they should be pursued.
- Such improvements are unlikely to be implemented without the continued involvement of independent citizen and scientist gadflies *and* a serious national commitment to a much reduced growth rate in our energy use.
- Finally, much more attention should be devoted to the issues associated with other parts of the nuclear fuel cycle—starting with the nuclear fuel reprocessing plant.

# Chernobyl

The disaster at the Chernobyl reactor shocked everybody. In the beginning, very little information was available, so many people in the United States felt that a similar disaster could happen here. Happily, as more information became available, it became clear that the design of the Chernobyl-type reactor differs totally from any commercial U.S. reactor. The lack of a containment building is only the least important of the differences. The internal design is far more important, and a lack of competence and discipline in the reactor's operation contributed to the disaster.

A special U.S. panel was formed to study the Chernobyl accident. The committee's preliminary report was issued in August 1986, just before the Soviets reported to the International Atomic Energy Agency in Vienna. The panel consisted of Frederick Seitz, Hans A. Bethe, Dixy Lee Ray, Miro M. Todorovich, Robert K. Adair, Bernard L. Cohen, Thomas J. Connolly, Herbert Goldstein, Herbert Kouts, Leon Lidofsky, John McCarthy, Eugene P. Wigner, and Edwin L. Zebroski.

The following is the main content of this panel's report. Some points that have become irrelevant in the meantime have been omitted, and some additional changes have been made. For the reader interested in the technical details, I recommend "Chernobyl: The Soviet Report," published in the October issue of *Nuclear News*.

The Chernobyl accident had several contributing factors, none of which would apply to a light-water reactor in the United States:

- The Chernobyl-type reactor design is seriously flawed. It involves intrinsic instability in operation. When it overheats, and steam forms in the cooling water, reactivity of the reactor increases, and the reactor tends to overheat even more. Therefore, under normal operation, these reactors are controlled by elaborate automatic computer control systems. Under the experimental conditions leading to the accident, the instability factor was increased so that it became autocatalytic (self-accelerating)—increasing steam output increased power sharply and a runaway situation resulted.

By contrast, in any light-water reactor, increases in the temperature or steam content of the water will cause the reactivity to go down. The reactor operation is inherently stable at low, partial, or full power.

- Errors in planning and conducting the experiment were so serious that they certainly indicate not only gross deficiencies in training and management, but also little appreciation of the unstable nature of the reactor. The personnel in charge of the Chernobyl-4 reactor made a series of serious mistakes, as has been described by the Soviets at the meeting of the International Atomic Energy Agency. They conducted an experiment that, in view of the unstable nature of the reactor, could and should have been done with the reactor shut down completely; they did not do so. Other very serious mistakes were made that made the reactor essentially uncontrollable. Apparently, not only the operating personnel but also the supervision and management, over a period of 12 hours, lacked appreciation of the reactor design characteristics that made it extremely dangerous to operate under the conditions they used.

By contrast, since the Three Mile Island accident, every U.S. reactor must have one or more fully trained nuclear engineers available at all times. On special occasions, such as the start-up of a new reactor, qualified engineers from the designer and supplier organizations as well as the Nuclear Regulatory Commission and the operating company are required to be present to review and approve procedures and to observe all details of the operation.

- The Chernobyl reactor, like all Soviet reactors of this type, and contrary to some earlier press reports, did *not* have a containment building over the reactor. (Some of the large pressurized-water-type reactors in the Soviet Union do have containment.) According to the Russian report: "A light cylindric housing (casing) enclosed the space of the graphite block structure." It serves to keep air away from the hot graphite. There is a shielding deck of thick concrete blocks to protect personnel against radiation, but portions are held down only by gravity. If several of the approximately 1,700 pressurized tubes in the reactor burst suddenly because of overheating, there is ample force and pressure available to rupture the thin cylindrical vessel and to lift sections of this deck, opening direct communication between the reactor compartment and the large reactor hall over the top of the reactor. The latter is an ordinary factory type structure, not designed to hold pressure. The overpressure in the hall when the reactor deck failed for whatever reasons immediately blew off most of the roof and large parts of the side walls.

Containment buildings around U.S. reactors are designed and tested to contain complete rapid release of water and steam, and also any radioactive materials released in case of an accident. They are also equipped with water sprays which cool down the steam and radioactive products, condensing them within the containment. If there is any escape of radioactive material from the containment building, it will consist essentially of the noble gases, which disperse rapidly and are biologically rather ineffective. All but traces of the other fission products will be retained inside the building.

Further comments on the comparison between the Chernobyl-type and light-water-type reactors are appropriate. For example: In a reactor accident, the radioactive fission products of greatest concern are iodine and cesium. Iodine is strongly combined with cesium as cesium iodide. Both iodine and cesium iodide are highly soluble in water, especially in slightly alkaline water normally used for containment-building sprays. For these reasons, Three Mile Island released to the environment about one-millionth of the amount of iodine that was ejected into the atmosphere at Chernobyl. No amounts of cesium or strontium were detected at Three Mile Island.

The likelihood of a seriously damaging accident is clearly much smaller for U.S. light-water reactors than for the Chernobyl-type owing to the inherently large safety margins of the U.S. design, in contrast to the inherent instability of the Chernobyl-type. Moreover, continuing process of open reporting and communication of deficiencies in operation—whether they be large or small—and their remedies, and continuing vigilance in implementing corrective actions, is one of the strengths of the U.S. system that give us confidence in making this distinction.

Further, there is clearly an additional very large factor of safety for the public for a light-water reactor with robust containment, even given that a core-damaging accident might again happen. In contrast, the essentially uncontained top of the Chernobyl reactor virtually insured that major environmental damage would occur if there were a core-damaging accident. The U.S.-type containment provides a large and important margin of safety for the public.

Both the accident at Chernobyl and the less serious one at Three Mile Island reflected complacency. While we understand that the safety margins in U.S. reactors are significantly larger than in the Chernobyl-type reactors, it is important to avoid complacency so that continued concern for safety is maintained over the full lifetime of U.S. plants.

# FIVE
# PHYSICISTS

# J. Robert Oppenheimer

J. Robert Oppenheimer did more than any other man to make American theoretical physics great.

His mind was concerned with the most fundamental questions in physics. His attitude of concentrating on the fundamental difficulties and ignoring the easy problems was communicated to his students. "What we don't understand we explain to each other," he once said in describing the activities of the physics group at the Institute for Advanced Studies at Princeton. There was always a burning question which had to be discussed from all aspects, a solution to be found, to be rejected, and another solution attempted. There was always life and excitement and the expectation of excitement in physics for generations to come.

Oppenheimer started in physics at the most opportune time, taking his B.A. at Harvard in 1925. In 1926, Schrödinger discovered his equation and that year Oppenheimer had written his Ph.D. thesis in Göttingen on an important application of that just-invented theory. He calculated the photoelectric effect in hydrogen and for X rays. Even today it is a complicated calculation, beyond the scope of most quantum-mechanics textbooks. Oppenheimer had to develop all the methods himself, including the normalization of wave functions in the continuum. Naturally, his calculations were later improved upon, but he correctly obtained the absorption coefficient at the $K$ edge and the frequency dependence in its neighborhood. He was disturbed by the fact that his theory, while agreeing well with measurements of X-ray absorption coefficients, did not seem in accord with the absorption of hydrogen in the sun. This, however, was the fault of limited understanding of the solar atmosphere in 1926—not of Oppenheimer's theory.

For four years, from 1925 to 1929, Oppenheimer traveled from one center of physics to another—Cambridge University and Göttingen as a Ph.D. student, Harvard and California Institute of Technology as a National Research Fellow, then Leyden and Zurich as a fellow of the International

Education Board. In Zurich he was influenced by Pauli, probably the man with the deepest understanding of quantum mechanics. In Göttingen, after completing his Ph.D., Oppenheimer worked with Max Born, one of the inventors of the then new quantum mechanics. Their paper on the structure of molecules is still the basis of our understanding of molecular spectra.

In 1929, Oppenheimer accepted a position as assistant professor at the University of California at Berkeley. Simultaneously, he held an appointment at the California Institute of Technology in Pasadena, where he regularly spent part of the year. This was the beginning of his great school of theoretical physics. In the fourteen years before Los Alamos, a large number of the best theoretical physicists in the United States, including Christy and Schiff, did their Ph.D. work with him. Soon his school became famous and attracted postdoctoral fellows like Serber and Schwinger. His lectures were a great experience, for experimental as well as theoretical physicists. In addition to a superb literary style, he brought a degree of sophistication in physics previously unknown in the United States. Here was a man who obviously understood all the deep secrets of quantum mechanics and who yet made it clear that the most important questions were unanswered. His earnestness and deep involvement gave his research students the same sense of challenge. He never gave his students easy and superficial answers but trained them to appreciate and work on deep problems. Many of them migrated with him between Berkeley and Pasadena every year.

The problems of nonrelativistic quantum mechanics had been pretty well solved by 1929. Now Dirac's relativistic wave equation of the electron became the great challenge. In 1930, Dirac advanced the hypothesis that the vexing negative-energy states in his equation were all normally occupied except for a few "holes," which he assumed corresponded to protons. Oppenheimer quickly showed that this last hypothesis was untenable, and that the holes must have the same mass as an electron. This conclusion led to the theoretical prediction of the positron, discovered two years later by Anderson in cosmic radiation, that great laboratory of nature which revealed to us so many new particles in the 1930s and 1940s.

Cosmic radiation was Millikan's chief interest. Millikan was then president of Caltec and its chief physicist. The very peculiar phenomenon, electron showers, had been observed, both in the atmosphere and in pieces of solid material. After the theory of the production of positrons and electrons (to which Oppenheimer and M. Plesset contributed the first paper) had been published, Oppenheimer and his school developed a most elegant theory of shower production which accounted for most of the observed phenomena and which has remained fundamentally unchanged. Other components of cosmic radiation were known to penetrate deep into the earth, recognized as mu mesons, after the discovery of that particle by Anderson and Neddermeyer, and these were known, in turn, to produce showers, although rarely. Oppenheimer's students Christy and Kusaka found this an indication that the meson had spin of 0 or ½. Particles of higher spin would give much too strong radiation.

At this point, cosmic-ray research tied in with Oppenheimer's other chief concern at the time — the fact that the theory gave divergent integrals for the self-energy and for the probability of certain processes at high energies. His struggle with this problem was intense, but he rejected all facile solutions. Concerning one theory by a prominent colleague which attempted to explain some showers of particularly rapid development, he said wryly, "What a shameless exploitation of divergent integrals." In the midst of these researches came the war, making a break in the lives of most American physicists, but in Oppenheimer's perhaps more than any other. After the war, the divergence of field theory and the internal contradictions of meson physics were still with us. As much as his official duties permitted, Oppenheimer returned to physics, which was entering a time of rapid and exciting development.

His influence on physics was greatly enhanced when, in 1947, he was offered the position of director of the Institute for Advanced Studies at Princeton. As head of its physics group, Oppenheimer realized, probably more fully than had ever been done before, the full possibilities of the Institute. Here was a place where dozens of the best and most active young theoretical physicists could assemble and could discuss the most interesting ideas of physics which kept streaming in faster than they could be digested. The Institute's physics department became the world's center for the development of high-energy physics and field theory. It is probably no exaggeration to say that, for the next ten years, it was the mecca of theoretical physics, as Copenhagen had been in the 1920s and 1930s.

Physics was now much more mature than it had been in the 1930s at Berkeley. So were the physicists who flocked to Princeton. They were all of postdoctoral status and many of them were of established prominence. Pauli was a frequent guest, and so were Dirac and Yukawa, who first proposed the theory of the meson. A large number of postdoctoral fellows received their final training and taste in physics at this great center. Among them were Gell-Mann, Goldberger, Chew, Low, Nambu, and others in this country, leaders in the development of modern theory. There were almost equally as many young visitors from abroad — France, Italy, England, Germany, and elsewhere. And then there was the superb, almost-permanent staff, including Dyson and Pais, as well as Lee and Yang, who did their revolutionary work on breaking of parity in weak interactions at the Institute.

Oppenheimer was always there to stimulate, to discuss, to listen to ideas. Even when he was busy with public affairs, he knew what was most important in physics. It was forever astonishing how quickly he could absorb new ideas and single out the most important point.

In 1948, I gave a seminar at the Institute on some calculations concerning the Lamb shift. I spoke for less than half the time; the rest of the time was given to discussion by the many bright young physicists, and especially by Oppenheimer himself. Ideas developed fast in this atmosphere of intense discussion and stimulation. Incidentally, I was told that I had been allowed to speak much longer than was customary.

Vigorous discussion as well as emphasis on fundamental problems was Oppenheimer's style. Perhaps this originated during his time at Göttingen in 1926, the formative year of quantum mechanics and of his scientific life; perhaps he wanted to perpetuate that feeling of continuous discovery which must have pervaded Göttingen. All through his life he was able to convey to all around him a sense of excitement in the quest of science.

He could also irritate the people who worked with him. His great mind was able to read and digest physics much faster than his less gifted colleagues. In scientific conversation, he always assumed that others knew as much as he. This being seldom the case, and few being willing to admit their ignorance, his partner often felt at a disadvantage. Yet, when asked directly, he explained willingly.

Aside from his work at the Institute, Oppenheimer played a leading role in high-energy conferences which brought together theoretical and experimental physicists annually. The first such meeting, organized by the Rockefeller Institute, was held in 1947 at Shelter Island. Some rather remarkable experimental results were presented—the Lamb shift and the anomalous magnetic moment of the electron. These results stimulated theorists to develop modern quantum electrodynamics and renormalization theory, which eliminated, to a large extent, the unpleasant divergences which had plagued prewar theory. Oppenheimer, most active at the first conference, organized the next two, giving to Schwinger and Feynman an opportunity to present their diverse solutions to this problem. Later, Marshak established a regular annual conference at Rochester, which soon became international and now is held alternately in Russia, Western Europe, and the United States. To the end, Oppenheimer was much involved in the organization of these meetings and was a regular participant.

To the world outside physics, Oppenheimer is best known as the director of the Los Alamos Scientific Laboratory during the war. I had the good fortune to participate in an activity preparatory to the work at Los Alamos. In the summer of 1942, a small group met under Oppenheimer's leadership to discuss theoretical methods of assembling an atomic weapon. By that time it was very likely that Fermi's atomic pile would work, that Dupont would build a production reactor, and that useful quantities of plutonium would be produced. The separation of uranium-235 by the electromagnetic method, although extremely expensive, also seemed very likely to succeed: separation by gaseous diffusion was less certain. In any case, the committee in charge of the uranium project considered it advisable to begin a serious study of the assembly of a weapon. It turned out to be accurate timing. Some members of our group, under the leadership of Serber, did calculations on the actual subject of our study, the neutron diffusion in an atomic bomb and the energy yield obtainable from it. The rest of us, especially Teller, Oppenheimer, and I, indulged ourselves in a far-off project — namely, the question of whether and how an atomic bomb could be used to trigger an H-bomb. Grim as the subject was, it was a most interesting enterprise. We were forever inventing new tricks, finding ways to calculate, and rejecting most of the tricks on the basis of the calculations. It was one of the best scientific collaborations I have ever experienced.

Life soon became more serious. After the summer study, we all went home to our respective tasks of war research, but in the fall plans were started which led to the founding of the Los Alamos Scientific Laboratory in March 1943. It was not at all clear that Oppenheimer would be its director. He had, after all, no experience in directing a large group of people. The laboratory would be devoted primarily to experiment and to engineering, and Oppenheimer was a theorist. It is greatly to the credit of General Groves, who by then was in charge of the "Manhattan Project," that he overruled all these objections and made Oppenheimer the director.

It was a marvelous choice. Los Alamos might have succeeded without him, but certainly only with much greater strain, less enthusiasm, and less speed. As it was, it was an unforgettable experience for all members of the laboratory. There were other wartime laboratories of high achievement, like the Metallurgical Laboratory at Chicago, the Radiation Laboratory at MIT, and others, here and abroad. But I have never observed in any of these other groups quite the spirit of belonging together, quite the urge to reminisce about the days of the laboratory, quite the feeling that this was really the great time of their lives.

That this was true of Los Alamos was mainly due to Oppenheimer. He was a leader. It was clear to all of us, whenever he spoke, that he knew everything that was important to know about the technical problems of the laboratory, and he somehow had it well organized in his head. But he was not domineering, he never dictated what should be done. He brought out the best in all of us, like a good host with his guests. And because he clearly did his job very well, in a manner all could see, we all strove to do our job as best we could.

One of the factors contributing to the success of the laboratory was its democratic organization. The governing board, where questions of general and technical laboratory policy were discussed, consisted of the division leaders (about eight of them). The coordinating council included all the group leaders, about 50 in number, and kept all of them informed on the most important technical progress and problems of the various groups in the laboratory. All scientists having a B.A. degree were admitted to the colloquium in which specialized talks about laboratory problems were given. Each of these three assemblies met once a week. In this manner, everybody in the laboratory felt a part of the whole and felt that he should contribute to the success of the program. Very often a problem discussed in one of these meetings would intrigue a scientist in a completely different branch of the laboratory and he would come up with an unexpected solution.

The free interchange of ideas was entirely contrary to the organization of the Manhattan District as a whole. As organized by General Groves, the work was strictly compartmentalized, with one laboratory having little or no knowledge of the problems or progress of the other. Oppenheimer had to fight hard for free discussion among all qualified members of the laboratory. But the free flow of information and discussion, together with Oppenheimer's personality, kept morale at its highest throughout the war.

As the war was coming to an end and the problem arose about what to do with atomic energy, the government appointed an interim committee to discuss it. Its members included Oppenheimer, other wartime laboratories participants of the Manhattan District, and several elder-statesmen scientists. One of the committee's meetings took place at Los Alamos and some other Los Alamos scientists were asked to participate. I remember it very vividly. All the participants were impressive people who had made great contributions. Nevertheless, whenever Oppenheimer left the room, discussion slid back into fairly routine problems, such as the specific nuclear reactions one should investigate and the kind of research that could be done with a nuclear reactor. On his return, the level of the discussion immediately rose and we all had the feeling that now the meeting had become really worthwhile.

With the end of the war, political problems came to the fore. Oppenheimer has often been blamed for his initial support of the May-Johnson bill, which provided for continued military control and severe penalties for any infraction of the rules. Oppenheimer supported it because he thought it was the only way to get atomic energy organized quickly. But he soon joined the mainstream of scientists and those in Congress supporting the McMahon bill, which in the end became law.

An even greater concern was the international treatment of atomic energy. During the war, Oppenheimer had listened carefully to Niels Bohr, who had very clear ideas about what an atomic armaments race would mean and had a plan to avoid it by making atomic energy international. Bohr had come to the United States in 1944 and had been asked to help us at Los Alamos. He was quite interested in our work and gave us some advice. However, his main interest was in talking to statesmen and trying to persuade them that international control of the atom was the only way to avoid a pernicious arms race or, worse, atomic war. Bohr did not succeed, but the combined efforts of statesmen and scientists after the war did result in some progress.

One result was the 1946 Acheson-Lilienthal Report. Oppenheimer played the leading role in the Lilienthal Committee. The report called for the creation of an international authority to control all atomic-energy work. The plan emphasized the need for a positive task for the international authority. It should develop atomic reactors for power and other peaceful uses, and also atomic weapons, if desired: it should not have merely the function of a policeman preventing individual nations from developing atomic energy and weapons on their own. This wise plan was endorsed by a State Department committee under Acheson and became official U.S. policy. It was presented to the United Nations by Baruch, but unfortunately was totally rejected by the U.S.S.R. Oppenheimer was one of the first to see that the plan would be rejected by Russia. Most of the members of the Federation of American Scientists held onto hope beyond hope. His realism, as well as his official duties, kept Oppenheimer rather separate from the Federation and other political organizations of scientists.

From 1947 to 1953, Oppenheimer was a familiar figure in Washington. His main function was that of chairman of the General Advisory Committee of the Atomic Energy Commission created in early 1947. But he also consulted with the Department of Defense on atomic weapons and on the general strategic policy of the United States. He was an important member of many *ad hoc* study groups on military matters. In all this he resisted, to the extent possible, the prevalent philosophy that atomic weapons give us "more bang for a buck." He, and others with him, advocated that more emphasis be put on atomic weapons for tactical use (so as to avoid a wholesale conflagration) and on conventional armament. This earned him the hostility of some elements of the Air Force.

The General Advisory Committee of the AEC was a group of extremely high-grade scientists and businessmen. In its early years, it recommended an extensive research effort by the AEC, which contributed greatly to the present preeminence of the U.S. in high-energy and nuclear physics. National laboratories like Brookhaven, Oak Ridge, and Argonne were established during this period, and the Berkeley Radiation Laboratory was strongly supported. In these years, the groundwork was laid for the development of nuclear power reactors by the AEC. The main task of the AEC and its General Advisory Committee was to ensure an ample supply of fissionable material for reactors, as well as atomic weapons, by constructing production facilities. Thanks to this effort we are now living in an age of atomic plenty.

The work of the General Advisory Committee came to a crisis in the fall of 1949, after the U.S.S.R. had exploded its first atomic weapon. In response, Edward Teller proposed that the U.S.should develop H-bombs. The committee wrote a strong recommendation against the development of the "super." One important argument was that there was, at that time, no sufficient technical basis for this development (the crucial invention was made in 1951 by Teller). Another strong argument was that the U.S. should not deliberately step up the arms race and should at least first make an effort to discuss with Soviet Russia the possibility of an agreement not to develop hydrogen weapons. This advice was overruled by President Truman, after several months of heated debate behind the scenes.

After President Truman had overruled the committee, it would probably have been right for Oppenheimer to resign as chairman. He tried to, but his resignation was not accepted. This fact, together with the hostility he had incurred in the Air Force for his opposition to strategic bombing, brought about his troubles in 1953 and 1954. They were introduced by a strange article in *Fortune* attacking him. In 1953, on the basis of a denunciation, President Eisenhower ordered that Oppenheimer's security clearance be terminated. The ensuing, long-protracted security investigation became a *cause célèbre*. Many of his scientist friends came to his defense. A few came out against him. The *Proceedings*, published by the AEC, give a vivid story of the discussions within the U.S. government on defense policy between 1947 and 1953. They have been avidly read by friend and foe abroad.

Both the Security Hearing Board, by a vote of 2 to 1, and the AEC, by a vote of 4 to 1, decided to withhold security clearance from Oppenheimer. In the final majority opinion by the Commission, the only real argument against granting him clearance was the grotesque story of Haakon Chevalier in 1942. Intrinsically this "espionage attempt" was of no importance whatever (the counterintelligence corps did not even bother to investigate the lead), but apparently Oppenheimer, under stress and overwork at Los Alamos, had invented a rather foolish cock-and-bull story to shield his friend, and had then denied it.

It was not until April 1962 that the government made amends. President Kennedy invited Oppenheimer to a White House dinner for Nobel prize winners. And in 1963, just after taking office, President Johnson gave Oppenheimer the highest honor given by the AEC, the $50,000 Fermi award. In his acceptance remarks, Oppenheimer said, "I think it is just possible, Mr. President, that it has taken some charity and some courage for you to make this award today."

Oppenheimer took the outcome of the security hearing very quietly, but he was a changed person; much of his previous spirit and liveliness had left him. Excluded from government work, he apparently did not have the strength to return to active work in physics. He was as interested and well-informed about physics as ever before, still a leading figure at international conferences. But his main activity was now along more general lines.

He was deeply concerned, before and after 1954, with the public understanding of science. His Reith lectures over the BBC, given in 1953 and published as *Science and the Common Understanding*, are among the most lucid and, at the same time, most profound popular expositions of atomic and quantum theory. Here, again, he never took the easy way of explaining just the facts, and he carefully avoided any facile analogies between the uncertainty principle and biological processes.

He was much aware of, and troubled by, the inability of the modern scientist to communicate his exhilarating experience of discovery, and also the contents of his discoveries, to the educated layman, in contrast to the close communication between science and society two centuries earlier [see, for example, "Some Reflections on Science and Culture" (1960)]. In still other lectures ["The Open Mind" (1955)], he discusses the relation of scientists to society and many facets of atomic policy of the United States. He always gives the impression of having long wrestled with the problem; he always raises a great many penetrating questions; and he gives few concrete answers.

If this left his audience only partly satisfied, they were compensated by the beauty of his style. I have seldom heard a speaker, scientist or otherwise, who had such command of the English language and who could fit words to the depth of the thought so well. There was wit also and a store of good anecdotes but, most of all, the signs of a deeply concerned human being.

Oppenheimer leaves a lasting memory with all scientists who worked with him and who passed through his school and whose taste in physics was formed by him. His was a truly brilliant mind, best described by his long-time associate Charles Lauritsen: "This man was unbelievable. He always gave you the answer before you had time to formulate the question."

# *Freeman Dyson*

E ducated in the best British classical tradition at home and at Win-
chester, endowed with an excellent memory and a thoughtful, as
well as sensitive, understanding of what he reads, Freeman Dyson
says the intuitive understanding of society and history by poets and other
writers is for him an ever-present guide to his own understanding. Thus, he
is able to carry us into his way of finding meaning through appropriate quo-
tations from poems and essays. One of his favorite thinkers is J. B. S.
Haldane, a biologist and poet. Dyson is a bridge between the two cultures
of C.P. Snow, the humanistic and the scientific-technical. He tries to show
non scientists what science and scientists are about.

The time of Dyson's entry into physics, the fall of 1947, was most fortunate. Willis Lamb and Polykarp Kusch had done their famous experiments
at Columbia, finding the Lamb shift and the anomalous magnetic moment
of the electron. I had published an approximate theory of the Lamb shift.
Dyson, coming to Cornell as a beginning graduate student, but having
received very advanced mathematical training in Cambridge, worked out
the relativistic theory of the Lamb shift for Bose particles, using old-
fashioned perturbation theory. More important for him and for physics, he
learned firsthand the methods of Richard Feynman, that is, using Feynman
graphs and the concept of positrons propagating backward in time. Feynman was then a professor at Cornell and loved to discuss his ideas with
Dyson. In the following summer, Dyson attended the Summer School at
Ann Arbor, in which Julian Schwinger explained in detail his methods of
relativistic quantum electrodynamics. Dyson was probably the only person
who thoroughly understood both methods. On a nonstop bus trip from the
West Coast back to the East on which he got very drowsy, he suddenly realized how to combine both methods. Dyson published a paper about this
synthesis after he had arrived at the Princeton Institute for Advanced Study.
In the same paper he proved that renormalization theory gives finite results

in all orders of perturbation theory. Of course, many people had to do a lot of work to implement Dyson's program.

At Princeton, Dyson discovered that Robert Oppenheimer, who had previously been much impressed by Schwinger's theory, now considered renormalization theory to be fundamentally on the wrong track, and that he would therefore be fighting for the entire program of quantum electrodynamics: "Already I could feel that the Lord had delivered him into my hands." The fight was more difficult than Dyson had anticipated. I was amused to read a description, quite accurate, of a seminar I gave at Princeton in which I hardly got a chance to say a few sentences. But Dyson reports that after this seminar, he himself was listened to, and finally convinced Oppenheimer.

In *Disturbing the Universe* (Harper & Row, 1979), Dyson gives a delightful picture of Feynman, with whom he shared a long drive from Ithaca to Albuquerque. Later comes a character sketch of Oppenheimer, perhaps the most perceptive I know. Dyson saw all his faults, but also his greatness.

Dyson says, "We are scientists second, human beings first. We become politically involved because knowledge implies responsibility. We fight as best we can for what we believe to be right. Often we fail." More than most of us, Dyson is concerned with the human condition. He is a young man to be writing an autobiography, and we soon discover that he views his life history as a convenient vehicle to talk about some of the fundamental issues of our time.

When Dyson was 20, he was a civilian scientist in the Operational Research Section of the British Bomber Command. This work was traumatic, especially when he found out that the bombing of German cities hardly affected the outcome of the war, while the losses among the British bomber crews, already high, were increasing, and that his section's analysis had no influence at all. It was a harsh encounter with death, entrenched bureaucracy, and the futility of much military action. Later, Dyson participated in two technological ventures he describes vividly. The first was TRIGA, an intrinsically safe nuclear reactor for research purposes. His (and a few other people's) theoretical invention was tested experimentally, and he had the great satisfaction that it actually worked as conceived. Many TRIGAs have been sold by the General Atomic Company, and the TRIGA may have been the only reactor that made money for its builder. The other technological venture was Orion, a device to propel a large spaceship into space by the recoil from many small nuclear explosions. Dyson and others working on the concept were extremely enthusiastic, and he describes happily the satisfaction of working together on an exciting problem with good results. However, funds were stopped and the project terminated.

In the early 1960s, Dyson worked extensively for the Arms Control and Disarmament Agency. He made important suggestions to ACDA which, however, were not acted on. Most importantly, he feels that when Kennedy and Khrushchev were leading the U.S. and the Soviet Union, it was an auspicious combination for a major reduction of armaments, not merely the test ban that they agreed upon. He regrets the waste of this opportunity in the brief span that these two statesmen were in power. The trouble is that is takes years for a leader to bring a country along with him, and this was true even for a magician of persuasion like Franklin Roosevelt.

Dyson would like military defense to become superior to offense. I would like very much to agree with him, but it is very hard to achieve this goal as long as nuclear weapons exist.

Dyson describes many emotional struggles and the growth of his self. His desire for clarity and his scientific training for truth enable him to analyze his motives and to recognize and so state that he was mistaken in specific thoughts and actions; examples are his initial opposition to the test ban and the account of four reasons why he was wrong; the analysis of why it was fortunate that the Orion project could not be continued.

One of Dyson's deep convictions is that there should be much diversity in the world. He strongly supports all the new independent nations, and also separate languages, like Welsh and Swiss German, which give their speakers a sense of community within a greater, sometimes oppressive entity. As an extreme measure for diversification, he wants to colonize space, a desire in which I cannot follow him. He recognizes that it will take a long time to realize this ambition, and that it may even require biological and not only cultural change.

He considers diversity essential also for technical inventions. For instance, he believes that nuclear power would be much better off, safer and more generally accepted, if many small groups of scientists could have invented different types of nuclear reactors. This is probably true, but in my opinion he underrates the need for detailed engineering, and for investment of large amounts of capital in establishing an industry.

Dyson is a very thoughtful man who has worked on many phases of science and its applications.

# Herman W. Hoerlin

## WITH DONALD M. KERR AND ROBERT A. JEFFRIES

Herman W. Hoerlin, widely recognized as one of the world's experts in the physics of high-altitude nuclear detonations was born on July 5, 1903, in Schwäbisch Hall, Germany and studied in Berlin and Stuttgart, where he received his Dr.Ing. in 1936. For his thesis, he measured cosmic rays incident upon the global region between Spitzbergen and the Magellan Strait and established the world's highest observation station, at 20,000 feet in the Peruvian Andes. Hoerlin confirmed the then much-disputed particle nature of cosmic rays within two months after Arthur H. Compton completed his worldwide studies and before Robert Millikan published his work.

Earlier, in 1930, Hoerlin was the chief photographer for an international Himalayan expedition and, with an Austrian friend, made the first ascent of Jongsong Peak, making it the highest mountain climbed to the summit. In 1938 he emigrated to the U.S. and joined the General Aniline and Film Corporation, where he became chief physicist.

With encouragement from Hans Bethe, Hoerlin moved to the Los Alamos National Laboratory in 1953. There he led the Nuclear Weapons Effect Group, participated in all atmospheric nuclear weapons tests and developed many pioneering diagnostic techniques and physical measurements. Because most of his research was related to defense, Hoerlin's technical excellence was not often demonstrated in the unclassified literature; however, his publications on high-altitude nuclear detonations are numerous and considered by most to be the definitive works. Twice he testified before Congress, once favoring an atmospheric test ban and later describing the technology involved in antiballistic missiles. From his retirement in 1982 until shortly before his death, Hoerlin continued to consult at Los Alamos not only in the field of weapons effects, but also on other technologies, such as inertial-confinement fusion.

Outside the defense community he maintained close contact with universities; in particular, as a visiting professor at Cornell University during 1959–1960 and as an active friend of St. John's College in Santa Fe. A fellow of APS, Hoerlin was a talented physicist and an effective administrator, a dedicated friend of nature and, in every respect, a true gentleman. Hoerlin died in Bedford, Massachusetts on November 6, 1983.

# Paul P. Ewald

## WITH H.J. JURETSCHKE, A.F. MOODIE, AND H.K. WAGENFELD

Paul P. Ewald, a key figure in the evolution of modern physics, was born in Berlin, Germany, in 1888. His father, who died shortly before Ewald was born, was a historian at the University of Berlin. His mother was an internationally successful portrait painter. He learned to speak English and French before starting school, and his classical education at the *Gymnasium* gave him a lifelong love of literature and languages, especially classical Greek; he readily quoted Homer throughout his life. Soon attracted to the sciences, he tried chemistry at Cambridge, then mathematics at Göttingen and Munich, where he was finally drawn to physics. He thought of Arnold Sommerfeld, David Hilbert and Alfred Pringsheim as his most important teachers. From among the thesis topics proposed to him, he chose what Sommerfeld considered the "least promising" one, namely "Dispersion and double refraction of electron lattices (crystals)." He had been fascinated since boyhood by light and its interaction with solid matter, and he was to be preoccupied by this topic for the rest of his life.

Shortly before submitting the thesis, he sought an interview with Max von Laue to clarify some details. In the course of this discussion, Laue conceived the idea of X-ray diffraction in crystals. Experiments quickly established X rays as waves and confirmed the existence of crystal lattices. This concept was basic to Ewald's thesis, which also contained most of the mathematical formalism of the dynamical theory of X-ray diffraction in perfect crystals. The theory was worked out fully shortly thereafter.

During his professorship in theoretical physics at the Technische Hochschule, Stuttgart (1921–1937), Ewald's department became an international center for X-ray diffraction and solid-state physics. In 1932 he became *Rektor* (equivalent to university president), but resigned this post soon after the Nazis came to power. He continued in his position as professor until 1937, when he was pensioned after walking out of a faculty meeting in protest over a speaker's statement: "Objectivity is no longer a valid or acceptable concept in science." Soon thereafter he left Germany for Cambridge, England, where he had been offered a small research grant. Subsequently, he held academic positions at Queen's University, Belfast, Northern Ireland (1939–1949), and the Polytechnic Institute of Brooklyn (now of New York (1949–1959), where as department head for seven years, he created a new center for research. Retirement in 1959 did not stop his research nor his many other endeavors relating to crystallography.

Ewald's celebrated theory of X-ray diffraction (1917) remains a masterpiece of a self-consistent theory of normal modes, including many-body interactions, and its treatment of optical boundary conditions at the microscopic level (the extinction theorem) is unsurpassed.

Forty years passed before the semiconductor industry produced crystals of sufficient perfection to permit detailed confirmation of his theoretical predictions. Today, this same theory enables industry to verify the vitally important perfection of its crystals, and in such applications as X-ray interferometry, leads to precise values for many solid-state parameters. The theory's influence, however, extends beyond X rays: The original theory of electron diffraction in crystals (by Hans Bethe) drew on Ewald's concepts, and its subsequent development as a quantitative tool was essentially dynamical. It may be surprising to learn that even the analysis of many of the contrast details in the electron microscope, which are generated by strain fields of dislocations, relies heavily on Ewald's conceptual description of two-beam dynamical interactions, including anomalous absorption (the Borrmann effect). An offshoot, the Ewald sum procedure, was originally invented to calculate the electrostatic energy of an ionic crystal and is widely used, for instance, in modern band-structure calculations.

For 60 years, Ewald was a prime mover in X-ray crystallography. His book *Kristalle und Röntgenstrahlen* (1923) gave the first comprehensive treatment of the subject, while *Fifty Years of X-Ray Diffraction* (1962) surveyed the mature field. Together with C. Hermann, he founded *Strukturbericht* (first published in 1931), a collection of results on crystal structures, which, with its successor volumes *Structure Reports*, is the standard structure-data repository for industry and science. The *International Tables*, also conceived by him in the 1930s, set the uniform nomenclature, units and standards for the specification of these data. After World War II, Ewald initiated *Acta Crystallographica* and acted as one of its chief editors from 1949 to 1959. He was president of the American Crystallographic Association for 1951–1952.

Ewald was very much concerned with the international character of his science. After World War II, he was instrumental in reestablishing the International Union of Pure and Applied Physics, serving as its first secretary-general and later as vice-president. He was instrumental in founding a separate International Union of Crystallography, of which he served as president for some time.

The success of these diverse and wide-ranging science-policy initiatives rested largely on Ewald's personal qualities as a scientist of high vision and standards, and as a diplomatic and convincing negotiator, but ultimately on his disarming honesty and modesty. To his students and colleagues he was not only a window into the larger world of science and the intellect, but also a model physicist in his scrupulous search for the physically correct and lucid formulation of ideas. He was equally demanding in his insistence, honed by long years as an editor, on the precise language in which these ideas were to be expressed. At the same time, he always enjoyed his science and was delighted when progress was made, either by himself or by others. He remained receptive to new ways of thinking and was ready to get them a fair hearing when normal processes for doing so broke down. Finally, there is no better testimony to his personal qualities— harmoniously complemented by those of his wife, Ella—than their enormous international circle of friends, who made the Ewald residence, wherever it was, an eagerly sought-out stopover point for travelers from far and wide.

In 1979 Ewald received the first Gregori Aminoff Medal of the Royal Swedish Academy, in honor of his lifelong accomplishments. In 1985, IUC established the Ewald Prize.

In his late eighties, Ewald told one of us that he would like to "finish his doctor's thesis" by finding a way to deduce the structure of a crystal directly from the intensities of the X-ray diffraction spots. He would have been delighted that the 1985 Nobel Prize in chemistry was awarded to Jerome Karle and Herbert Hauptman for finding a solution to this problem. Ewald died at his home in Ithaca, New York on August 22, 1985 at the age of 97.

# Richard P. Feynman

Richard P. Feynman was one of the greatest physicists since the Second World War and, I believe, the most original. He won the Nobel prize for physics in 1965 for his work on quantum electrodynamics (QED). In 1948, simultaneously with Schwinger and Tomonaga (with whom he shared the prize), he showed that QED can be renormalized, removing infinities, so that it gives finite results for all observable quantities. Schwinger and Tomonaga did this by building on existing theories in an ingenious way. Feynman invented a completely new method of treating the Schrödinger (or Dirac) equation. This was typical of Feynman's physics: he always had his own way of looking at and solving a problem.

Both methods were essential: Schwinger's immediately convinced quantum theorists, because it connected with previous knowledge. Feynman's seemed abstruse at first and Niels Bohr, for instance, found it hard to accept, but it soon became clear that it enormously simplified concepts and calculations. Only by these methods was it possible to extend theory to increasingly complicated problems and quantum electrodynamics to fabulous accuracy: the magnetic moment of the electron has now been calculated to an accuracy of one part in $10^8$, and measured to a similar accuracy.

Feynman also developed, in 1953, a fundamental theory of liquid helium, justifying the earlier theories of Landau and Tisza. Because $^4$He atoms are Bose particles, having spin zero, the ground-state wave function of a mass of liquid helium is symmetrical in all the particles and everywhere positive. There is only one such wave function, and the entire mass of helium behaves as one unit. This is why helium near 0 K is superfluid, showing no viscosity. At low temperature, pressure waves are the only possible motions in the liquid. At a higher temperature, about .5 K, it becomes possible for a small ring of atoms to circulate without other atoms being much disturbed; these are the "rotons" of Landau theory. Feynman showed why the energy of a roton is minimum at a certain wavelength and that this wavelength is closely related to the average distance of helium atoms from each other.

To make a roton takes energy, therefore their number increases with temperature. They also interact with each other and thus show viscosity. An assembly of rotons therefore behaves much like a normal liquid, and moves independently of the superfluid. Helium may be regarded as a mixture of superfluid and normal liquid. When the concentration of normal liquid becomes too big, there is a phase transition in which the whole liquid turns "normal."

Feynman's work on the nuclear weak interaction is also of fundamental importance. In 1956, Lee and Yang concluded that parity symmetry is not conserved in weak interactions, and this was soon confirmed experimentally by Wu and others. On the basis of experiments, physicists concluded that the violation of parity is the maximum possible. This stimulated Feynman and Gell-Mann to postulate that only the left-handed part of the wave function of a particle is involved in weak interactions. They postulated, in fact, that this was true not only for the neutrino, but for any particle, electron, muon, and even composite particles like protons and neutrons. They also proposed that the weak interaction is universal: all particles interact and with the same strength. This theory permitted many conclusions, nearly all of them agreeing with experiments. One experiment at first seemed to disagree, but when repeated more carefully, agreed with the Feynman–Gell-Mann theory. The concept of chirality, left- or right-handed spin, turns out to be extremely fruitful in particle theory.

Experiments in the late 1960s on the scattering of high-energy electrons by protons at the Stanford Linear Accelerator showed very large inelastic scattering. Feynman soon concluded that smaller units were contained in the protons which he called partons, and that these collided elastically with electrons. The partons were soon identified with the quarks deduced from general theory, and one of the important tasks of particle physics has been the determination of the distribution of quarks in protons and neutrons. Feynman's attempts to understand quantum chromodynamics, the theory of parton interactions, were characteristically individual, especially his treatment of the confinement of quarks in hadrons. But his long struggle with abdominal cancer cut this work short.

Feynman enjoyed all life, and he lived in physics. He was able to transmit his enthusiasm to others, his colleagues and his students. He was beloved by my children, then two to six years old, for whom he babysat many times while he was at Cornell. It helped that he was also a clown. Once, when he was explaining nuclear fission at a high school, he did it simultaneously to two classes, standing in the doorway drawing pictures on two blackboards simultaneously with his right and left hands.

Feynman's approach was always direct, to life as to science. He disliked pompous people and made fun of them, but usually gently so as not to hurt them. His uncanny ability to get immediately to the core of a problem became well known when he sat on the commission to investigate the Challenger disaster and demonstrated the central problem simply by dropping a rubber O-ring into a glass of ice water.

In physics he always stayed close to experiment, and had no interest in esoteric theories. The three volumes, *The Feynman Lectures on Physics,* which tie the elementary problems of physics with the most advanced ideas, exemplify this. More than other scientists, he was loved by his colleagues and his students. Feynman died February 15, 1988.

# ASTROPHYSICS

# Energy Production in Stars

From time immemorial, people must have been curious to know what keeps the sun shining. The first scientific attempt at an explanation was by Helmholtz about 100 years ago, and was based on the force most familiar to physicists at the time, gravitation. When a gram of matter falls to the sun's surface it gets a potential energy,

$$E_{pot} = -GM/R = -1.91 \times 10^{15} \text{ erg/g} \tag{1}$$

where $M = 1.99 \times 10^{33}$ g is the sun's mass, $R = 6.96 \times 10^{10}$ cm its radius, and $G = 6.67 \times 10^{-8}$ the gravitational constant. A similar energy was set free when the sun was assembled from intersellar gas or dust in the dim past; actually somewhat more, because most of the sun's material is located closer to its center, and therefore has a numerically larger potential energy. One-half of the energy set free is transformed into kinetic energy, according to the well-known virial theorem of mechanics. This will permit us later to estimate the temperature in the sun. The other half of the potential energy is radiated away. We know that at present the sun radiates

$$\epsilon = 1.96 \text{ erg/g sec.} \tag{2}$$

Therefore, if gravitation supplies the energy, there is enough energy available to supply the radiation for about $10^{15}$ sec which is about 30 million years.

This was long enough for nineteenth-century physicists, and certainly a great deal longer than man's recorded history. It was not long enough for the biologists of the time. Darwin's theory of evolution had just become popular, and biologists argued with Helmholtz that evolution would require a longer time than 30 million years, and that therefore his energy source for the sun was insufficient. They were right.

At the end of the nineteenth century, radioactivity was discovered by

Becquerel and two Curies who received one of the first Nobel prizes for this discovery. Radioactivity permitted a determination of the age of the earth, and more recently, of meteorites which indicate the time at which matter in the solar system solidified. On the basis of such measurements, the age of the sun is estimated to be 5 milliards of years, within about 10 percent [1 milliard $= 10^9$]. So gravitation is not sufficient to supply its energy over the ages.

Eddington, in the 1920s, investigated very thoroughly the interior constitution of the sun and other stars, and was very concerned about the sources of stellar energy. His favorite hypothesis was the complete annihilation of matter, changing nuclei and electrons into radiation. The energy which was to be set free by such a process, if it could occur, is given by the Einstein relation between mass and energy and is

$$c^2 = 9 \times 10^{20} \text{ erg/g.} \tag{3}$$

This would be enough to supply the sun's radiation for 1,500 milliards of years. However, nobody has ever observed the complete annihilation of matter. From experiments on earth, we know that protons and electrons do not annihilate each other in $10^{30}$ years. It is hard to believe that the situation would be different at a temperature of some 10 million degrees such as prevails in the stars, and Eddington appreciated this difficulty quite well.

From the early 1930s it was generally assumed that the stellar energy is produced by nuclear reactions. Already in 1929, Atkinson and Houtermans[1] concluded that at the high temperatures in the interior of a star, the nuclei in the star could penetrate into other nuclei and cause nuclear reactions, releasing energy. In 1933, particle accelerators began to operate in which such nuclear reactions were actually observed. They were found to obey very closely the theory of Gamow, Condon, and Gurney, on the penetration of charged particles through potential barriers. In early 1938, Gamow and Teller[2] revised the theory of Atkinson and Houtermans on the rate of "thermonuclear" reactions—nuclear reactions occurring at high temperature. At the same time, Weizsäcker[3] speculated on the reactions which actually might take place in the stars.

In April 1938, Gamow assembled a small conference of physicists and astrophysicists in Washington, D.C., sponsored by the Department of Terrestrial Magnetism of the Carnegie Institution. At the conference, the astrophysicists told us physicists what they knew about the internal constitution of the stars. This was quite a lot, and all their results had been derived without knowledge of the specific source of energy. The only assumption they made was that most of the energy was produced "near" the center of the star.

The most easily observable properties of a star are its total luminosity and its surface temperature. In relatively few cases of nearby stars, the mass of the star can also be determined.

Figure 1 shows the customary Hertzsprung-Russell diagram. The luminosity, expressed in terms of that of the sun, is plotted against the surface temperature, both on a logarithmic scale. Conspicuous is the main sequence, going from upper left to lower right—from hot and luminous stars to cool and faint ones. Most stars lie on this sequence. In the upper right are the red giants, cool but brilliant stars. In the lower left are the white dwarfs, hot but faint. We shall be mainly concerned with the main sequence. After being assembled by gravitation, stars spend the most part of their life on the main sequence, then develop into red giants, and in the end, probably into white dwarfs. The figure shows that typical surface temperatures are of the order of $10^4$ K.

Figure 2 gives the relation between mass and luminosity in the main sequence. At the upper end, beyond about 15 sun masses, the mass determinations are uncertain. It is clear, however, that luminosity increases rapidly with mass. For a factor of 10 in mass, the luminosity increases by a factor of about 3,000, hence the energy production per gram is about 300 times larger.

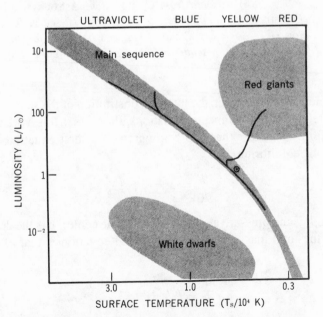

FIG. 1. *Hertzsprung-Russell Diagram. From E.E. Salpeter, in* Apollo and the Universe *(Science Foundation for Physics, University of Sydney, 1967).*

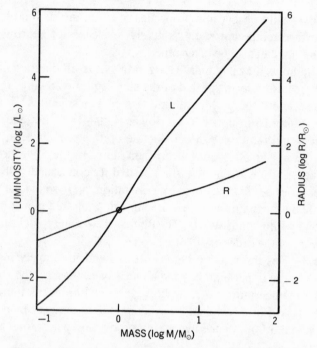

Fɪɢ. 2. *Luminosity and radius of stars vs. mass. Abscissa is log* M/M$_\odot$. *Data from C. W. Allen,* Astrophysical Quantities *(Athlone Press, 1963), p. 203. The curve for* log L/L$_\odot$ *holds for all stars, that for* R/R$_\odot$ *only for the stars in the main sequence. The symbol* $_\odot$ *refers to the sun.*

To obtain information on the interior constitution of the stars, astrophysicists integrate two fundamental equations. Pioneers in this work have been Eddington, Chandrasekhar and Strömgren. The first equation is that of hydrostatic equilibrium

$$\frac{dP}{dr} = -GM(r)\,\frac{\rho(r)}{r^2} \tag{4}$$

in which $P$ is the pressure at distance $r$ from the center, $\rho$ is the density and $M(r)$ is the total mass inside $r$. The second equation is that of radiation transport

$$\frac{1}{\chi\rho}\frac{d}{dr}\left(\tfrac{1}{3}acT^4\right) = -\frac{L(r)}{4\pi r^2}. \tag{5}$$

Here $\chi$ is the opacity of the stellar material for blackbody radiation of the local temperature $T$, $a$ is the Stefan-Boltzmann constant, and $L(r)$ is the flux of radiation at $r$. The value of $L$ at the surface $R$ of the star is the luminosity. In the stars we shall discuss, the gas obeys the equation of state

$$P = RT\rho/\mu \qquad (6)$$

where $R$ is the gas constant, while $\mu$ is the mean molecular weight of the stellar material. If $X$, $Y$, and $Z$ are respectively concentrations by mass of hydrogen, helium and all heavier elements, and if all gases are fully ionized, then

$$\mu^{-1} = 2X + 3Y/4 + Z/2. \qquad (7)$$

In all stars, except the very oldest ones, it is believed that $Z$ is between .02 and .04; in the sun at present, $X$ is about .65, hence, $Y = .33$ and $\mu = .65$. In many stars, the chemical composition, especially $X$ and $Y$, vary with position $r$. The opacity is a complicated function of $Z$ and $T$, but in many cases it behaves like

$$\chi = C\rho T^{-3.5} \qquad (8)$$

where $C$ is a constant.

The integration of (4) and (5) in general requires computers. However, an estimate of the central temperature may be made from the virial theorem which we mentioned at the beginning. According to this theorem, the average thermal energy per unit mass of the star is one-half of the average potential energy. This leads to the estimate of the thermal energy per particle at the center of the star

$$kT_c = \alpha\mu GHM/R \qquad (9)$$

in which $H$ is the mass of the hydrogen atom, and $\alpha$ is a constant whose magnitude depends on the specific model of the star but is usually about 1 for main sequence stars. Using this value, and (1), we find for the central temperature of the sun

$$T_{6c} = 14 \qquad (10)$$

where $T_6$ denotes the temperature in millions of degrees, here and in the following. A more careful integration of the equations of equilibrium by Demarque and Percy[4] gives

$$T_{6c} = 15.7; \quad \rho_c = 158 \text{ g/cm}^3. \tag{11}$$

Originally, Eddington had assumed that the stars contain mainly heavy elements, from carbon on up. In this case $\mu = 2$ and the central temperature is increased by a factor of 3, to about 40 million degrees; this understanding led to contradictions with the equation of radiation flow (5), if the theoretical value of the opacity was used. Strömgren pointed out that these contradictions can be resolved by assuming the star to consist mainly of hydrogen, which is also in agreement with stellar spectra. In modern calculations, the three quantities $X$, $Y$, $Z$, indicating the chemical composition of the star, are taken to be parameters to be fixed so as to fit all equations of stellar equilibrium.

All nuclei in a normal star are positively charged. In order for them to react they must penetrate each others' Coulomb potential barrier. The wave mechanical theory of this shows that in the absence of resonances, the cross section has the form

$$\sigma(E) = \frac{S(E)}{E} \exp\left(-\sqrt{\frac{E_G}{E}}\right) \tag{12}$$

where $E$ is the energy of the relative motion of the two colliding particles, $S(E)$ is a coefficient characteristic of the nuclear reaction involved and

$$E_G = 2M(\pi Z_0 Z_1 e^2/h)^2 = (2\pi Z_0 Z_1)^2 E_{\text{Bohr}}. \tag{13}$$

Here $M$ is the reduced mass of the two particles, $Z_0$ and $Z_1$ their charges, and $E_{\text{Bohr}}$ is the Bohr energy for mass $M$ and charge 1. Equation (13) can be evaluated to give

$$E_G = 0.979W \text{ MeV} \tag{14}$$

with
$$W = AZ_0^2 Z_1^2, \tag{14a}$$
$$A = A_0 A_1/(A_0 + A_1) \tag{14b}$$

in which $A_0$, $A_1$ are the atomic weights of the two colliding particles. For most nuclear reactions, $S(E)$ is between 10 MeV barns and 1 keV barn.

The gas at a given $r$ in the star has a given temperature so that the particles have a Boltzmann energy distribution. The rate of nuclear reactions is then proportional to

$$(8/\pi M)^{1/2}(kT)^{-3/2} \int \sigma(E) \, E \, \exp\left(-E/kT\right) dE. \tag{15}$$

It is most convenient[5] to write, for the rate of disappearance of one of the reactants,

$$dX_0/dt = - [01] X_0 X_1 \tag{16}$$

where $X_0$ and $X_1$ are the concentrations of the reactants by mass, and

$$[01] = 7.8 \times 10^{11} (Z_0 Z_1/A)^{1/3} S_{eff} \rho T_6^{-2/3} e^{-r}, \tag{17}$$
$$r = 42.487 (W/T_6)^{1/3}. \tag{17a}$$

Since the reaction cross section (12) increases rapidly with energy, the main contribution to the reaction comes from particles which have an energy many times the average thermal energy. Indeed the most important energy is

$$E_0 = (r/3) kT. \tag{18}$$

For $T = 13$, which is an average for the interior of the sun, we have

$$
\begin{aligned}
r/3 = \quad & 4.7 \text{ for the reaction H + H} \\
& 19 \quad \text{for the reaction C + H} \\
& 25 \quad \text{for the reaction N + H.}
\end{aligned} \tag{19}
$$

It is also easy to see from (17) that the temperature dependence of the reaction rate is

$$\frac{d \log[01]}{d \log T} = \frac{r - 2}{3}. \tag{20}$$

Evidently, at a given temperature and under otherwise equal conditions, the reactions which can occur most easily are those which have the smallest possible value of $W$, Eq. (14a). This means that at least one of the interacting nuclei should be a proton, $A_0 = Z_0 = 1$. Thus we may examine the reactions involving protons.

The simplest of all possible reactions is

$$H + H = D + \epsilon^+ + \nu \tag{21}$$

($\epsilon^+$ = positron, $\nu$ = neutrino). This conclusion was first suggested by Weizsäcker[3], and calculated by Critchfield and Bethe.[6] The reaction is exceedingly slow of course because it involves the beta disintegration.

Indeed, the characteristic factor $S$ is

$$S(E) = 3.36 \times 10^{-25} \text{ MeV barns.} \qquad (22)$$

This equation has been derived on purely theoretical grounds, using the known coupling constant of beta disintegration; the value is believed to be accurate to 5 percent or better. There is no chance of observing such a slow reaction on earth, but in the stars we have almost unlimited time, and a large supply of protons of high energy. As we shall see presently, the rate of energy production by this simple reaction fits the observed energy production in the sun very well.

The deuterons formed in (21) will quickly react further, and the end product is $He^4$. We shall discuss the reactions in more detail later on.

The proton-proton reaction (21), although it predicts the correct energy production in the sun, has a rather weak dependence on temperature. According to (19), (20), it behaves about as $T^4$. Since central temperatures change only little from the sun to more massive stars, the energy production by this reaction does likewise. However, as we have seen in Figure 2, the observed energy production increases dramatically with increasing mass. Therefore, there must exist nuclear reactions which are more strongly dependent on temperature; these must involve heavier nuclei.

Stimulated by the April 1938 Washington Conference, and following the argument just mentioned, I examined[7] the reactions between protons and other nuclei, going up in the periodic system. Reactions between H and $He^4$ lead nowhere, there being no stable nucleus of mass 5. Reactions of H with Li, Be and B, as well as with deuterons, are all very fast at the central temperature of the sun, but just this speed of the reaction rules them out: The partner of H is very quickly used up in the process. In fact, and just because of this reason, all the elements mentioned, from deuterium to boron, are extremely rare on earth and in the stars, and can therefore not be important sources of energy.

The next element, carbon, behaves quite differently. In the first place, it is an abundant element, probably making up about 1 percent by mass of any newly formed star. Secondly, in a gas of stellar temperature, it undergoes a cycle of reactions, as follows

$$C^{12} + H = N^{13} + \gamma, \qquad (23a)$$
$$N^{13} = C^{13} + \epsilon^+ + \nu, \qquad (23b)$$
$$C^{13} + H = N^{14} + \gamma, \qquad (23c)$$

$$N^{14} + H = O^{15} + \gamma, \qquad (23d)$$
$$O^{15} = N^{15} + \epsilon^+ + \nu, \qquad (23e)$$
$$N^{15} + H = C^{12} + He^4. \qquad (23f)$$

Reactions (23a), (23c), and (23d) are radiative captures; the proton is captured by the nucleus and the energy emitted in the form of gamma rays; these are then quickly converted into thermal energy of the gas. For reactions of this type, $S(E)$ is of the order of 1 keV barn. Reactions (23b) and (23e) are simply spontaneous beta decays, with lifetimes of 10 and 2 minutes respectively, negligible in comparison with stellar times. Reaction (23f) is the most common type of nuclear reaction, with 2 nuclei resulting from the collision; $S(E)$ for such reactions is commonly of the order of MeV barns.

Reaction (23f) is in a way the most interesting because it closes the cycle: We reproduce the $C^{12}$ which we started from. In other words, carbon is only used as a catalyst; the result of the reaction is a combination of four protons and two electrons[8] to form one $He^4$ nucleus. In this process two neutrinos are emitted, taking away about 2 MeV energy together. The rest of the energy, about 25 MeV per cycle, is released usefully to keep the sun warm.

Making reasonable assumptions of the reaction strength $S(E)$, on the basis of general nuclear physics, I found in 1938 that the carbon-nitrogen cycle gives about the correct energy production in the sun. Since it involves nuclei of relatively high charge, it has a strong temperature dependence, as given in (19). The reaction with $N^{14}$ is the slowest of the cycle and therefore determines the rate of energy production; it goes about as $T^{24}$ near solar temperature. This is amply sufficient to explain the high rate of energy production in massive stars.[9]

To put the theory on a firm basis, it is important to determine the strength factor $S(E)$ for each reaction by experiment. This has been done under the leadership of W. A. Fowler[10] of the California Institute of Technology in a monumental series of papers extending over a quarter of a century. Not only have all the reactions in (23) been observed, but in all cases $S(E)$ has been accurately determined.

The main difficulty in this work is due to the resonances which commonly occur in nuclear reactions. Figure 3 shows the cross section of the first reaction[5] (23a), as a function of energy. The measured cross sections extend over a factor of $10^7$ in magnitude; the smallest ones are $10^{-11}$ barns $= 10^{-35}$ cm$^2$ and therefore clearly very difficult to observe. The curve shows a resonance at 460 keV. The solid curve is determined from nuclear

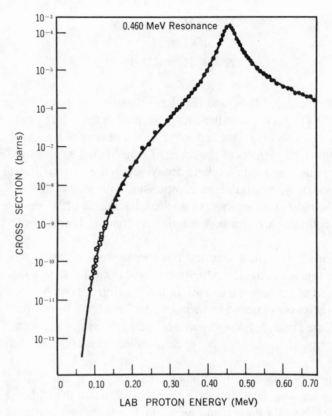

Fig. 3. *Cross section for the reaction $C^{12}$ + H, as a function of the proton energy. From Fowler, Caughlan and Zimmerman (reference 5).*

reaction theory, on the basis of the existence of that resonance. The fit of the observed points to the calculated curve is impressive. Similar results have been obtained on the other three proton-capture reactions in (23).

On the basis of Figure 3, we can confidently extrapolate the measurements to lower energy. As we mentioned in (18), the most important energy contributing to the reaction rate is about $20kT$. For $T_6 = 13$, we have $kT = 1.1$ keV; so we are most interested in the cross section around 20 keV. This is much too low an energy to observe the cross section in the laboratory; even at 100 keV, the cross section is barely observable. So quite a long extrapolation is required. This can be done with confidence provided there are no resonances close to $E = 0$. Therefore a great deal of experimental work has gone into the search for such resonances.

The resonances exist of course in the compound nucleus—the nucleus obtained by adding the two initial reactants. To find resonances near the threshold of the reactions (23), it is necessary to produce the same compound nucleus from other initial nuclei, e.g., in the reaction between $N^{14}$ and H, the compound nucleus $O^{15}$ is formed. To investigate its levels, Hensley[11] at the California Institute of Technology studied the reaction

$$O^{16} + He^3 = O^{15} + He^4. \qquad (24)$$

He found indeed a resonance 20 KeV below the threshold for $N^{14} + H$ which in principle might enhance the process (23d). However, the state in $O^{15}$ was found to have a spin $J = 7/2$. Therefore, even though $N^{14}$ has $J = 1$ and the proton has a spin of 1/2, we need at least an orbital momentum $\lambda = 2$, to reach this resonant state in $O^{15}$. The cross section for such a high orbital momentum is reduced by at least a factor $10^4$, compared to $\lambda = 0$, so that the near-resonance does not in fact enhance the cross section $N^{14} + H$ appreciably. This cross section can then be calculated by theoretical extrapolation from the measured range of proton energies, and the same is true for the other reactions in the cycle (23).

On this basis, Fowler and others have calculated the rate of reactions in the CN cycle. A convenient tabulation has been given by Reeves[12]; his results are plotted in Figure 4 which gives the energy production per gram per second as a function of temperature. We have assumed $X = .5$, $Z = .02$. The figure shows that at low temperature, the H + H reaction

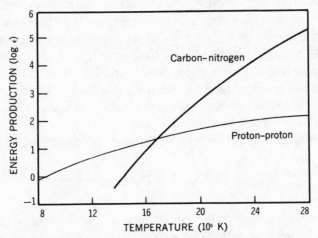

FIG. 4. *The energy production in erg/g sec, as a function of the temperature in millions of degrees, for the proton-proton reaction (P P) and the carbon-nitrogen cycle (C N). Concentrations assumed X = Y = .5, Z = .02. Calculated from Tables 8 and 9 of Reeves (reference 12).*

dominates, at high temperatures the C + N cycle; the crossing point is at $T_6 = 13$; here the energy production is 7 erg/g sec. The average over the entire sun is obviously smaller, and the result is compatible with an average production of 2 erg/g sec.

The energy production in the main sequence can thus be considered as well understood.

An additional point should be mentioned. Especially at higher temperature, when the CN cycle prevails, there is also a substantial probability for the reaction chain

$$O^{16} + H = F^{17} + \gamma, \tag{25a}$$
$$F^{17} = O^{17} + \epsilon^+ + \nu, \tag{25b}$$
$$O^{17} + H = N^{14} + He^4. \tag{25c}$$

This chain is not cyclic but feeds into the CN cycle. It is customary to speak of the whole set of reactions as the CNO bi-cycle. The effect of reactions (25) is that $O^{16}$ initially present will also contribute to the reactants available, and thus increase the reaction rate of CN cycle somewhat. This has been taken into account in Figure 4.

If equilibrium is established in the CNO bi-cycle, eventually most of the nuclei involved will end up as $N^{14}$ because this nucleus has by far the longest lifetime against nuclear reactions. There is no observable evidence for this behavior; in fact, wherever the abundance can be observed, C and O tend to be at least as abundant as N. However, this is probably owing to the fact that the interior of a star stays well separated from its surface; there is very little mixing. Astrophysicists have investigated the circumstances when mixing is to be expected, and have found that surface abundances are quite compatible with these expectations. In the interstellar material which is used to form stars, we have reason to believe that C and O are abundant and N is rare. This will be discussed later.

The initial reaction (21) is followed almost immediately by

$$H^2 + H = He^3 + \gamma. \tag{26}$$

The fate of $He^3$ depends on the temperature. Below about $T_6 = 15$, the $He^3$ builds up sufficiently so that two such nuclei react with each other according to

$$2He^3 = He^4 + 2H. \tag{27}$$

This reaction has an unusually high $S(E) = 5$ MeV barns.[5] At higher temperature the reaction

$$He^4 + He^3 = Be^7 + \gamma \qquad (28)$$

competes favorably with (27). The $Be^7$ thus formed may again react in one of two ways

$$Be^7 + e^- = 2He^4 + \nu, \qquad (29a)$$
$$Be^7 + H = B^8 + \gamma, \qquad (29b)$$
$$B^8 = 2He^4 + \epsilon^+ + \nu. \qquad (29c)$$

At about $T_6 = 20$, reaction (29b) begins to dominate over (29a). Equation (29b) is followed by (29c) which emits neutrinos of very high energy. Davis,[13] at Brookhaven, is attempting to observe these neutrinos.

A main sequence star uses up its hydrogen preferentially near its center where nuclear reactions proceed most rapidly. After a while, the center loses almost all its hydrogen. For stars of about twice the luminosity of the sun, this effect happens in less than $10^{10}$ years, which is approximately the age of the universe, and also the age of stars in the globular clusters. We shall now discuss what happens to a star after it has used up its hydrogen at the center. Of course, in the outside regions hydrogen is still abundant.

This evolution of a star was first calculated by Schwarzschild[14] who has been followed by many others; we shall use recent calculations by Iben.[15] When hydrogen gets depleted, not enough energy is produced near the center to sustain the pressure of the outside layers of the star. Hence, gravitation will cause the center to collapse. Thereby, higher temperatures and densities are achieved. The temperature also increases farther out where hydrogen is still left, and this region now begins to burn. After a relatively short time, a shell of H, away from the center, produces most of the energy; this shell gradually moves outward and gets progressively thinner as time goes on.

At the same time, the region of the star outside the burning shell expands. This result follows clearly from many numerical computations on this subject. The physical reason is not clear. One hypothesis is that it is due to the discontinuity in mean molecular weight: Inside the shell, there is mostly helium, of $\mu = 4/3$, outside we have mostly hydrogen, and $\mu = .65$. Another suggestion is that the flow of radiation is made difficult by the small radius of the energy source, and that this has to be compensated by lower density just outside the source.

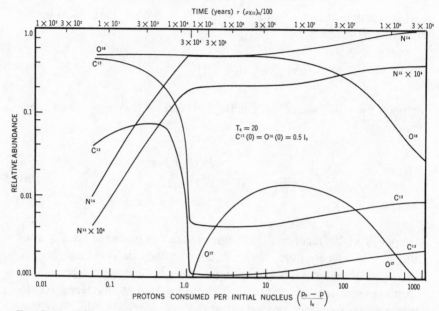

FIG. 5. *Variation with time of the abundances of various elements involved in the CNO cycle. It is assumed that initially $C^{12}$ and $O^{16}$ have the same abundance while that of $N^{14}$ is small. From G. R. Caughlan,* Astrophys. J. *(1967).*

By this expansion, the star develops into a red giant. Indeed, in globular clusters (which, as I mentioned, are made up of very old stars), all the more luminous stars are red giants. In the outer portion of these stars, radiative transport is no longer sufficient to carry the energy flow: therefore convection of material sets in in these outer regions. This convection can occupy as much as the outer 80 percent of the mass of the star; it leads to intimate mixing of the material in the convection zone.

Iben[15] has discussed a nice observational confirmation of this convectional mixing. The star Capella is a double star, each component having a mass of about three solar masses, and each being a red giant. The somewhat lighter star, Capella F (its spectral type if F) shows noticeable amounts of Li in its spectrum, while the somewhat heavier Capella G shows at least 100 times less Li. It should be expected that G, being heavier, is farther advanced in its evolution. Iben now gives arguments that the deep-reaching convection and mixing which we just discussed will occur just between the evolution phases F and G. By convection, material from the interior of the star will be carried to the surface; this material has been very hot and has therefore burned up its Li. Before deep convection sets in (in star F) the surface Li never achieves high temperatures and thus is preserved.

Following Iben's calculations, we have plotted in Figures 6 to 9 the development of various important quantities in the history of a star of mass = three solar masses. The time is in units of $10^8$ years. Since the developments go at very variable speed, the time scale has been broken twice, at $t = 2.31$ and $t = 2.55$. In between is the period during which the shell source develops.

During this period, the central temperature rises spectacularly (Figure 6), from about $T_6 = 25$ to $T_6 = 100$. At the same time, the radius increases from about 2 to 30 solar radii; subsequently, it decreases again to about 15 (Figure 7). The central density, starting at about 40, increases in the same period to about $5 \times 10^4$. The luminosity (Figure 9) does not change spectacularly, staying always between 100 and 300 times that of the sun.

While the inside and the outside of the star undergo such spectacular changes, the shell in which the hydrogen is actually burning does not change very much. Figure 9 shows $m$, the fraction of the mass of the star enclosed by the burning shell. Even at the end of the calculation, $t = 3.25$, this is only $m = .2$. This means that only 20 percent of the hydrogen in the star has burned after all this development. Figure 6, curve $T_s$, shows the

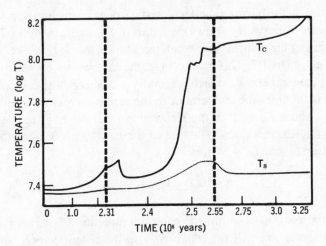

FIG. 6. *Evolution of a star of 3 solar masses, according to I. Iben, Astrophys. J. 142, 1447 (1965). Abscissa is time in units of $10^8$ years (note the breaks in scale at $t = 2.31$ and 2.55). I. Temperature (on logarithmic scale): $T_c$ = temperature at center of star, $T_s$ = same at midpoint of source of energy generation, which after $t = 2.48$ is a thin shell. $T_c$ increases enormously, $T_s$ stays almost constant.*

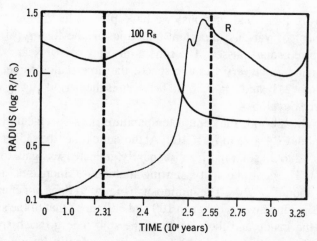

Fig. 7. *Evolution of a star (see caption to figure 6). II. Radius, in units of that of the sun, on logarithmic scale. R = total radius, $100\,R_s$ = 100 times the radius of midpoint of energy source. R increases tremendously, while $R_s$ shrinks somewhat.*

temperature in the burning shell which stays near 25 million degrees all the time. Figure 7, curve $R_s$, shows the radius of the shell, in units of the solar radius; during the critical time when the shell is formed, this radius drops from about .15 to .07. This is, of course, the mechanism by which the shell is kept at the temperature which originally prevailed at the center.

In the meantime, the temperature at the center increases steadily. When it reaches about $T_6 = 100$, the He$^4$ which is abundant at the center, can undergo nuclear reactions. The first of these, which occurs at the lowest temperature (about $T_6 = 90$), is

$$N^{14} + He^4 = F^{18} + \gamma. \tag{30}$$

While this reaction goes on, the central temperature remains fairly constant. However, there is not much N$^{14}$, so the reaction soon stops (after about $0.02 \times 10^8$ years), and the center contracts further.

The next reaction makes use entirely of the abundant He$^4$, viz.

$$3He^4 = C^{12} + \gamma. \tag{31}$$

This reaction has the handicap of requiring a simultaneous collision of three alpha particles. This would be extremely unlikely were it not for the fact that it is favored by a *double* resonance. Two alpha particles have nearly the same energy as the unstable nucleus $Be^8$, and further $Be^8 + He^4$ has almost the same energy as an excited state of $C^{12}$. This reaction can of course not be observed in the laboratory but the two contributing resonances can be. The importance of the first resonance was first suggested by Salpeter,[16] the second by Hoyle.[17] Recent data indicate that (31) requires a temperature of about $T_6 = 110$, at the central densities corresponding to $t = 2.5$, i.e., $\rho_e > 10^4$. Once this reaction sets in, the central temperature does not rise very fast anymore.

Reaction (31) is most important for the buildup of elements. Early investigators[3,7] had great trouble with bridging the gap between $He^4$ and $C^{12}$. Two nuclei in this gap, mass 5 and mass 8, are completely unstable, the rest disintegrate in a very short time under stellar conditions. Reaction (31), however, leads to stable $C^{12}$. This nucleus can now capture a further alpha particle

$$C^{12} + He^4 = O^{16} + \gamma. \tag{32}$$

The temperatures required for this are about the same as for (31). There is also some capture of alpha particles by $O^{16}$ leading to $Ne^{20}$, but the next step, $Ne^{23} \to Mg^{24}$, cannot occur appreciably at these temperatures; instead, the helium gets used up in forming $C^{12}$, $O^{16}$, and some $Ne^{20}$.

Helium is depleted first in the center, and now the same process repeats

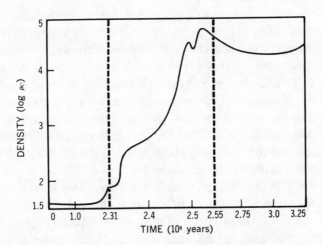

Fig. 8. *Evolution of a star (see caption to figure 6). III. Density, on logarithmic scale, at the center of the star. This quantity increases about 1000-fold.*

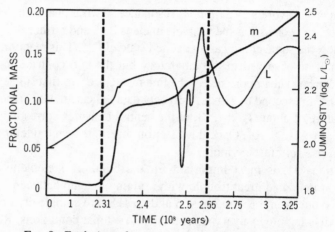

FIG. 9. *Evolution of a star (see caption to figure 6). IV. Curve L, luminosity relative to that of the sun, on logarithmic scale. This quantity does not change very much during the life of the star. Curve m, fraction of the mass of the star enclosed by energy-producing shell, on linear scale. This fraction increases slowly with time.*

which previously took place with hydrogen. A shell of burning He is formed, at a smaller radius than the H shell, and of course at a higher temperature. The center of the star now contracts further by gravitation and reaches still higher temperatures.

The further developments of a massive star are more speculative. However, the theory of Hoyle and collaborators[18] is likely to be correct.

The center of the star heats up until the newly formed carbon nuclei can react with each other. This happens at a temperature of roughly $10^9$ degrees. Nuclei like $Mg^{24}$ or $Si^{28}$ can be formed. There are also more complicated mechanisms in which we first have a capture reaction with emission of a gamma ray, followed by capture of this gamma ray in another nucleus which releases $He^4$. This $He^4$ can then enter further nuclei and build up the entire chain of stable nuclei up to the most stable Fe. Not much energy is released in all of these processes.

The center of the star contracts further and gets still hotter. At very high temperatures, several milliards of degrees, thermal equilibrium is no longer strongly in favor of nuclei of the greatest binding energy. Instead, endothermic processes can take place which destroy some of the stable nuclei already formed. In the process, alpha particles, protons, and even neutrons may be released. This permits the buildup of elements beyond Fe, up to the top of the periodic table. Because of the high temperatures involved all this probably goes fairly fast, perhaps in thousands of years.

During this stage, nuclear processes tend to consume rather than release energy. Therefore they no longer oppose the gravitational contraction so that contraction continues unchecked. It is believed that this will lead to an unstable situation. Just as the first contraction, at the formation of the H shell source, led to an expansion of the outer envelope of the star, a similar outward expansion is expected now. But time scales are now short, and this expansion may easily be an explosion. Hoyle *et al.* have suggested this as the mechanism for a supernova.

In a supernova explosion, much of the material of the star is ejected into interstellar space. We see this, for example, in the Crab Nebula. The ejected material probably contains the heavy elements which have been formed in the interior of the massive star. Thus, heavy elements get into the interstellar gas, and can then be collected again by newly forming stars. It is believed that this is the way stars get their heavy elements. This means that most of the stars we see, including our sun, are at least second generation stars, which have collected the debris of earlier stars which have suffered a supernova explosion.

To clinch this argument it must be shown that heavy elements cannot be produced in other ways. This has indeed been shown by Fowler.[19] He has investigated the behavior of the enormous gas cloud involved in the original big bang and its development with time. He has shown that temperatures and densities, as functions of time, are such that heavy elements beginning with C cannot be produced. The only element which can be produced in the big bang is $He^4$.

If all this is true, stars have a life cycle much like animals. They get born, they grow, they go through a definite internal development, and finally they die, to give back the material of which they are made so that new stars may live.

I am very grateful to Professor E. E. Salpeter for his extensive help in preparing this essay.

## References

1. R. d'E. Atkinson, F. G. Houtermans, *Zeits. f. Phys.* **54,** 656 (1929).

2. G. Gamow, E. Teller, *Phys. Rev.* **53,** 608 (1938).

3. C. F. von Weizsäcker, *Phys. Zeits.* **38,** 176 (1937).

4. P. R. Demarque, J. R. Percy, *Astrophys. J.* **140,** 541 (1964).

5. W. A. Fowler, G. R. Caughlan, B. A. Zimmerman, *Annual Review of Astronomy and Astrophysics* (Palo Alto, CA) **5,** 525 (1967).

6. H. A. Bethe, C. L. Critchfield, *Phys. Rev.* **54,** 248 (1938).

7. H. A. Bethe, *Phys. Rev.* **55,** 436 (1939).

8. The electrons are used to annihilate the positions emitted in reactions (23b) and (23e).

9. The carbon–nitrogen cycle was also discovered independently by C. F. von Weizsäcker, *Phys. Zeits.* **39,** 633 (1938), who recognized that this cycle consumes only the most abundant element, hydrogen. But he did not investigate the rate of energy production or its temperature dependence.

10. W. A. Fowler, many papers in *Phys. Rev.*, *Astrophys. J.* and other publications. Some of this work is summarized in reference 5.

11. D. C. Hensley, *Astrophys. J.* **147,** 818 (1967).

12. H. Reeves, in *Stellar Structure* (vol. 8 of *Stars and Stellar Systems*. G. P. Kuiper, editor, U. of Chicago Press, 1965), especially Tables 8 and 9.

13. R. Davis, Jr., *Phys. Rev. Lett.* **12,** 303 (1964).

14. M. Schwarzschild, *Structure and Evolution of the Stars* (Princeton Univ. Press, 1958).

15. I. Iben, Jr. *Astrophys. J.* **141,** 993 (1965); **142,** 1447 (1965); **143,** 483 (1966).

16. E. E. Salpeter, *Phys. Rev.* **88,** 547 (1952).

17. F. Hoyle, *Astrophys. J. Suppl.* **1,** 121 (1954).

18. E. M. Burbidge, G. R. Burbidge, W. A. Fowler, F. Hoyle, *Rev. Mod. Phys.* **29,** 547 (1957).

19. W. A. Fowler, International Assoc. of Geochemistry and Cosmochemistry, first meeting, Paris 1967.

# How a Supernova Explodes

## WITH GERALD BROWN

The death of a large star is a sudden and violent event. The star evolves peacefully for millions of years, passing through various stages of development, but when it runs out of nuclear fuel, it collapses under its own weight in less than a second. The most important events in the collapse are over in milliseconds. What follows is a supernova, a prodigious explosion more powerful than any since the big bang with which the universe began.

A single exploding star can shine brighter than an entire galaxy of several billion stars. In the course of a few months it can give off as much light as the sun emits in a billion years. Furthermore, light and other forms of electromagnetic radiation represent only a small fraction of the total energy of the supernova. The kinetic energy of the exploding matter is 10 times greater. Still more energy—perhaps 100 times more than the electromagnetic emission—is carried away by the massless particles called neutrinos, most of which are emitted in a flash that lasts for about a second. When the explosion is over, most of the star's mass has been scattered into space, and all that remains at the center is a dense, dark cinder. In some cases even that may disappear into a black hole.

Such an outline description of a supernova could have been given almost 30 years ago, and yet the detailed sequence of events within the dying star is still not known with any certainty. The basic question is this: A supernova begins as a collapse, or implosion; how does it come about, then, that a major part of the star's mass is expelled? At some point the inward movement of stellar material must be stopped and then reversed; an *implosion* must be transformed into an *explosion*.

Through a combination of computer simulation and theoretical analysis a coherent view of the supernova mechanism is beginning to emerge. It appears the crucial event in the turnaround is the formation of a shock wave that travels outward at 30,000 kilometers per second or more.

Supernovas are rare events. In our own galaxy, just three have been recorded in the past 1,000 years; the brightest of these, noted by Chinese observers in the year 1054, gave rise to the expanding shell of gas now known as the Crab Nebula. If only such nearby events could be observed, little would be known about supernovas. Because they are so luminous, however, they can be detected even in distant galaxies, and 10 or more per year are now sighted by astronomers.

The first systematic observations of distant supernovas were made in the 1930s by Fritz Zwicky of the California Institute of Technology. About half of the supernovas Zwicky studied fitted a quite consistent pattern: the luminosity increased steadily for about three weeks and then declined gradually over a period of six months or more. He designated the explosions in this group type I. The remaining supernovas were more varied, and Zwicky divided them into four groups; today, however, they are all grouped together as type II. In type I and type II supernovas the events leading up to the explosion are thought to be quite different. Here we shall be concerned primarily with type II supernovas.

The basis for the theory of supernova explosions was the work of Fred Hoyle of the University of Cambridge. The theory was then developed in a fundamental paper published in 1957 by E. Margaret Burbidge, Geoffrey R. Burbidge and William A. Fowler, all of Caltech, and Hoyle. They proposed that when a massive star reaches the end of its life, the stellar core collapses under the force of its own gravitation. The energy set free by the collapse expels most of the star's mass, distributing the chemical elements formed in the course of its evolution throughout interstellar space. The collapsed core leaves behind a dense remnant, in many cases a neutron star.

A supernova is an unusual and spectacular outcome of the sequence of nuclear fusion reactions that is the life history of a star. The heat given off by the fusion creates pressure, which counteracts the gravitational attraction that would otherwise make the star collapse. The first series of fusion reactions have the net effect of welding four atoms of hydrogen into a single atom of helium. The process is energetically favorable: the mass of the helium atom is slightly less than the combined masses of the four hydrogen atoms, and the energy equivalent of the excess mass is released as heat.

The process continues in the core of the star until the hydrogen there is used up. The core then contracts, since gravitation is no longer opposed by energy production, and as a result both the core and the surrounding material are heated. Hydrogen fusion then begins in the surrounding layers. Meanwhile, the core becomes hot enough to ignite other fusion reactions, burning helium to form carbon, then burning the carbon to form neon, oxygen and finally silicon. Again each of these reactions leads to the release of energy. One last cycle of fusion combines silicon nuclei to form iron, specifically the common iron isotope $^{56}$Fe, made up of 26 protons and 30 neutrons. Iron is the end of the line for spontaneous fusion. The $^{56}$Fe nucleus is the most strongly bound of all nuclei, and further fusion would absorb energy rather than releasing it.

At this stage in the star's existence, it has an onionlike structure. A core of iron and related elements is surrounded by a shell of silicon and sulfur, and beyond this are shells of oxygen, carbon and helium. The outer envelope is mostly hydrogen.

Only the largest stars proceed all the way to the final, iron-core stage of the evolutionary sequence. A star the size of the sun gets no further than helium burning, and the smallest stars stop with hydrogen fusion. A larger star also consumes its stock of fuel much sooner, even though there is more of it to begin with; because the internal pressure and temperature are higher in a large star, the fuel burns faster. Whereas the sun should have a lifetime of 10 billion years, a star 10 times as massive can complete its evolution 1,000 times faster. Regardless of how long it takes, all the usable fuel in the core will eventually be exhausted. At that point, heat production in the core ends and the star must contract.

When fusion ends in a small star, the star slowly shrinks, becoming a white dwarf: a burned-out star that emits only a faint glow of radiation. In isolation the white dwarf can remain in this state indefinitely, cooling gradually but otherwise changing little. What stops the star from contracting further? The answer was given more than 50 years ago by Subrahmanyan Chandrasekhar of the University of Chicago.

Loosely speaking, when ordinary matter is compressed, higher density is achieved by squeezing out the empty space between atoms. In the core of a white dwarf this process has reached its limit: the atomic electrons are pressed tightly together. Under these conditions the electrons offer powerful resistance to further compression.

Chandrasekhar showed there is a limit to how much pressure can be resisted by the electrons' mutual repulsion. As the star contracts, the gravitational energy increases, but so does the energy of the electrons, raising their pressure. If the contraction goes very far, both the gravitational energy and the electron energy are inversely proportional to the star's radius. Whether or not there is some radius at which the two opposing forces are in balance, however, depends on the mass of the star. Equilibrium is possible only if the mass is less than a critical value, now called the Chandrasekhar mass. If the mass is greater than the Chandrasekhar limit, the star must collapse.

The value of the Chandrasekhar mass depends on the relative numbers of electrons and nucleons (protons and neutrons considered collectively): the higher the proportion of electrons, the larger the electron pressure and so the larger the Chandrasekhar mass. In small stars where the chain of fusion reactions stops at carbon the ratio is approximately 1/2 and the Chandrasekhar mass is 1.44 solar masses. This is the maximum stable mass for a white dwarf.

A white dwarf with a mass under the Chandrasekhar limit can remain stable indefinitely; nevertheless, it is just such stars that are thought to give rise to type I supernovas. How can this be? The key to the explanation is that white dwarfs that explode in supernovas are not solitary stars but rather are members of binary star systems. According to one hypothesis, matter from the binary companion is attracted by the intense gravitational field of the dwarf star and gradually falls onto its surface, increasing the mass of the carbon-and-oxygen core. Eventually the carbon ignites at the center and burns in a wave that travels outward, destroying the star.

The idea that explosive carbon burning triggers type I supernovas was proposed in 1960 by Hoyle and Fowler. More detailed models have since been devised by many astrophysicists, most notably Icko Iben, Jr., and his colleagues at the University of Illinois at Urbana-Champaign. Recent calculations done by Ken'ichi Nomoto and his colleagues at the University of Tokyo suggest that the burning is actually not explosive. The wave of fusion reactions propagates like the burning of a fuse rather than like the explosion of a firecracker; it is a deflagration rather than a detonation.

Even though the burning is less violent than a detonation, the white dwarf is completely disrupted. The initial binding energy that holds the star together is approximately $10^{50}$ ergs; the energy released by the burning is 20 times greater ($2 \times 10^{51}$ ergs), enough to account for the 10,000-kilometer-per-second velocity of supernova remnants. In the course of the deflagration nuclear reactions create about one solar mass of the unstable nickel isotope $^{56}Ni$, which decays into $^{56}Co$ and then $^{56}Fe$ over a period of months. The rate of energy release from the radioactive decay is just right to account for the gradually declining light emission from type I supernovas.

The type II supernovas that are our main concern here arise from much more massive stars. The lower limit is now thought to be about eight solar masses.

In tracing the history of a type II supernova it is best to begin at the moment when the fusion of silicon nuclei to form iron first becomes possible at the center of the star. At this point, the star has already passed through stages of burning hydrogen, helium, neon, carbon and oxygen, and it has the onionlike structure described above. The star has taken several million years to reach this state. Subsequent events are much faster.

When the final fusion reaction begins, a core made up of iron and a few related elements begins to form at the center of the star, within a shell of silicon. Fusion continues at the boundary between the iron core and the silicon shell, steadily adding mass to the core. Within the core, however, there is no longer any production of energy by nuclear reactions; the core is an inert sphere under great pressure. It is thus in the same predicament as a white dwarf: it can resist contraction only by electron pressure, which is subject to the Chandrasekhar limit.

Once the fusion of silicon nuclei begins, it proceeds at an extremely high rate, and the mass of the core reaches the Chandrasekhar limit in about a day. We noted above that for a white dwarf the Chandrasekhar mass is equal to 1.44 solar masses; for the iron core of a large star the value may be somewhat different, but it is probably in the range between 1.2 and 1.5 solar masses.

When the Chandrasekhar mass has been attained, the pace speeds up still more. The core that was built in a day collapses in less than a second. The task of analysis also becomes harder at this point, so that theory relies on the assistance of computer simulation. Computer programs that trace the evolution of a star have been developed by a number of workers, including W. David Arnett of the University of Chicago and a group at the Lawrence Livermore National Laboratory led by Thomas A. Weaver of that laboratory and Stanford Woosley of the University of California at Santa Cruz. They are the "burners" of stars; we and our colleagues in theoretical physics are "users" of their calculations.

The simulations furnish us with a profile of the presupernova core, giving composition, density and temperature as a function of radius. The subsequent analysis relies on applying familiar laws of thermodynamics, the same laws that describe such ordinary terrestrial phenomena as the working of a heat engine or the circulation of the atmosphere.

I t is worthwhile tracing in some detail the initial stages in the implosion of the core. One of the first points of note is that compression raises the temperature of the core, which might be expected to raise the pressure and slow the collapse. Actually the heating has just the opposite effect.

Pressure is determined by two factors: the number of particles in a system and their average energy. In the core both nuclei and electrons contribute to the pressure, but the electron component is much larger. When the core is heated, a small fraction of the iron nuclei are broken up into smaller nuclei, increasing the number of nuclear particles and raising the nuclear component of the pressure. At the same time, however, the dissociation of the nuclei absorbs energy; since energy is released when an iron nucleus is formed, the same quantity of energy must be supplied in order to break the nucleus apart. The energy comes from the electrons and decreases their pressure. The loss in electron pressure is more important than the gain in nuclear pressure. The net result is that the collapse accelerates.

It might seem that the implosion of a star would be a chaotic process, but in fact it is quite orderly. Indeed, the entire evolution of the star is toward a condition of greater order, or lower entropy. It is easy to see why. In a hydrogen star each nucleon can move willy-nilly along its own trajectory, but in an iron core groups of 56 nucleons are bound together and must move in lockstep. Initially the entropy per nucleon, expressed in units of Boltzmann's constant, is about 15; in the presupernova core it is less than 1. The difference in entropy has been carried off during the evolution of the star by electromagnetic radiation and toward the end also by neutrinos.

The low entropy of the core is maintained throughout the collapse. Nuclear reactions continually change the species of nuclei present, which one might think could lead to an increase in entropy; the reactions are so fast, however, that equilibrium is always maintained. The collapse takes only milliseconds, but the time scale of the nuclear reactions is typically from $10^{-15}$ to $10^{-23}$ second, so that any departure from equilibrium is immediately corrected.

Another effect was once thought to increase the entropy, but it now seems likely that it actually reduces it somewhat. The high density in the collapsing core favors the reaction known as electron capture. In this process a proton and an electron come together to yield a neutron and a neutrino. The neutrino escapes from the star, carrying off both energy and entropy and cooling the system just as the evaporation of moisture cools the body. There are several complications to this process, so that its effect on the entropy is uncertain. In any case, the loss of the electron diminishes the electron pressure and so allows the implosion to accelerate further.

The first stage in the collapse of a supernova comes to an end when the density of the stellar core reaches a value of about $4 \times 10^{11}$ grams per cubic centimeter. This is by no means the maximum density, since the core continues to contract, but it marks a crucial change in physical properties: at this density matter becomes opaque to neutrinos. The importance of this development was first pointed out by T. J. Mazurek of the Mission Research Laboratory in Santa Barbara, California, and by Katsushiko Sato of the University of Tokyo.

The neutrino is an aloof particle that seldom interacts with other forms of matter. Most of the neutrinos that strike the earth, for example, pass all the way through it without once colliding with another particle. When the density exceeds 400 billion grams per cubic centimeter, however, the particles of matter are packed so tightly that even a neutrino is likely to run into one. As a result neutrinos emitted in the collapsing core are effectively trapped there. The trapping is not permanent; after a neutrino has been scattered, absorbed and reemitted many times, it must eventually escape, but the process takes longer than the remaining stages of the collapse. The effective trapping of neutrinos means that no energy can get out of the core.

The process of electron capture in the early part of the collapse reduces not only the electron pressure but also the ratio of electrons to nucleons, the quantity that figures in the calculation of the Chandrasekhar mass. In a typical presupernova core the ratio is between .42 and .46; by the time of neutrino trapping it has fallen to .39. This lower ratio yields a Chandrasekhar mass of .88 solar mass, appreciably less than the original value of between 1.2 and 1.5.

At this point the role of the Chandrasekhar mass in the analysis of the supernova also changes. At the outset it was the largest mass that could be supported by electron pressure; it now becomes the largest mass that can collapse as a unit. Areas within this part of the core can communicate with one another by means of sound waves and pressure waves, so that any variations in density are immediately evened out. As a result the inner part of the core collapses homologously, or all in one piece, preserving its shape.

The theory of homologous collapse was worked out by Peter Goldreich and Steven Weber of Caltech, and was further developed by Amos Yahil and James M. Lattimer of the State University of New York at Stony Brook. The shock wave that blows off the outer layers of the star forms at the edge of the homologous core. Before we can give an account of that process, however, we must continue to trace the sequence of events within the core itself.

Chandrasekhar's work showed that electron pressure cannot save the core of a large star from collapse. The only other hope for stopping the contraction is the resistance of nucleons to compression. In the presupernova core nucleon pressure is a negligible fraction of electron pressure. Even at a density of $4 \times 10^{11}$ grams per cubic centimeter, where neutrino trapping begins, nucleon pressure is insignificant. The reason is the low entropy of the system. At any given temperature, pressure is proportional to the number of particles per unit volume, regardless of the size of the individual particles. An iron nucleus, with 56 nucleons, makes the same contribution to the pressure as an isolated proton does. If the nuclei in the core were broken up, their pressure might be enough to stop the contraction. The fissioning of the nuclei is not possible, however, because the entropy of the core is too low. A supernova core made up of independently moving protons and neutrons would have an entropy per nucleon of between five and eight, whereas the actual entropy is less than one.

The situation does not change, and the collapse is not impeded, until the density in the central part of the core reaches about $2.7 \times 10^{14}$ grams per cubic centimeter. This is the density of matter inside a large atomic nucleus, and in effect the nucleons in the core merge to form a single gigantic nucleus. A teaspoonful of such matter has about the same mass as all the buildings in Manhattan combined.

Nuclear matter is highly incompressible. Hence once the central part of the core reaches nuclear density there is powerful resistance to further compression. That resistance is the primary source of the shock waves that turn a stellar collapse into a spectacular explosion.

Within the homologously collapsing part of the core, the velocity of the infalling material is directly proportional to distance from the center. (It is just this property that makes the collapse homologous.) Density, on the other hand, decreases with distance from the center, and as a result so does the speed of sound. The radius at which the speed of sound is equal to the infall velocity is called the sonic point, and it marks the boundary of the homologous core. A disturbance inside the core can have no influence beyond this radius. At the sonic point sound waves move outward at the speed of sound, as measured in the coordinate system of the infalling matter. This matter is moving inward at the same speed, however, and so the waves are at a standstill in relation to the center of the star.

When the center of the core reaches nuclear density, it is brought to rest with a jolt. This gives rise to sound waves that propagate back through the medium of the core, rather like the vibrations in the handle of a hammer when it strikes an anvil. The waves slow as they move out through the homologous core, both because the local speed of sound declines and because they are moving upstream against a flow that gets steadily faster. At the sonic point they stop entirely. Meanwhile additional material is falling onto the hard sphere of nuclear matter in the center, generating more waves. For a fraction of a millisecond the waves collect at the sonic point, building up pressure there. The bump in pressure slows the material falling through the sonic point, creating a discontinuity in velocity. Such a discontinuous change in velocity constitutes a shock wave.

At the surface of the hard sphere in the heart of the star infalling material stops suddenly but not instantaneously. The compressibility of nuclear matter is low but not zero, and so momentum carries the collapse beyond the point of equilibrium, compressing the central core to a density even higher than that of an atomic nucleus. We call this point the instant of "maximum scrunch." Most computer simulations suggest the highest density attained is some 50 percent greater than the equilibrium density of a nucleus. After the maximum scrunch the sphere of nuclear matter bounces back, like a rubber ball that has been compressed. The bounce sets off still more sound waves, which join the growing shock wave at the sonic point.

A shock wave differs from a sound wave in two respects. First, a sound wave causes no permanent change in its medium; when the wave has passed, the material is restored to its former state. The passage of a shock wave can induce large changes in density, pressure and entropy. Second, a sound wave—by definition—moves at the speed of sound. A shock wave moves faster, at a speed determined by the energy of the wave. Hence once the pressure discontinuity at the sonic point has built up into a shock wave, it is no longer pinned in place by the infalling matter. The wave can continue outward, into the overlying strata of the star. According to computer simulations, it does so with great speed, between 30,000 and 50,000 kilometers per second.

Up to this point in the progress of the supernova essentially all calculations are in agreement. What happens next, however, is not yet firmly established. In the simplest scenario, which we have favored, the shock wave rushes outward, reaching the surface of the iron core in a fraction of a second and then continuing through the successive onionlike layers of the star. After some days it works its way to the surface and erupts as a violent explosion. Beyond a certain radius—the bifurcation point—all the material of the star is blown off. What is left inside the bifurcation radius condenses into a neutron star.

Alas! Using presupernova cores simulated in 1974 by Weaver and Woosley, calculations of the fate of the shock wave are not so accommodating. The shock travels outward to a distance of between 100 and 200 kilometers from the center of the star, but then it becomes stalled, staying at roughly the same position as matter continues to fall through it. The main reason for the stalling is that the shock breaks up nuclei into individual nucleons. Although this process increases the number of particles, which might be expected to raise the pressure, it also consumes a great deal of energy; the net result is that both temperature and pressure are sharply reduced.

The fragmentation of the nuclei contributes to energy dissipation in another way as well: it releases free protons, which readily capture electrons. The neutrinos emitted in this process can escape, removing their energy from the star. The escape is possible because the shock has entered material whose density is below the critical value for neutrino trapping. The neutrinos that had been trapped behind the shock also stream out, carrying away still more energy. Because of the many hazards to the shock wave in the region between 100 and 200 kilometers, we have named this region of the star the "minefield."

It would be satisfying to report that we have found a single mechanism capable of explaining for all type II supernovas how the shock wave makes its way through the minefield. We cannot do so. What we have to offer instead is a set of possible explanations, each of which seems to work for stars in a particular range of masses.

The place to begin is with stars of between 12 and about 18 solar masses. Weaver and Woosley's most recent models of presupernova cores for such stars differ somewhat from those they calculated a decade ago; the most important difference is that the iron core is smaller than earlier estimates indicated—about 1.35 solar masses. The homologous core, at whose surface the shock wave forms, includes .8 solar mass of this material, leaving .55 solar mass of iron outside the sonic point. Since the breaking up of iron nuclei has the highest energy cost, reducing the quantity of iron makes it easier for the shock to break out of the core.

Jerry Cooperstein and Edward A. Baron of Stony Brook have been able to simulate successful supernova explosions in computer calculations that begin with Weaver and Woosley's model cores. The main requirement, first surmised by Sidney H. Kahana of the Brookhaven National Laboratory, is that the homologous core be very strongly compressed, so that it can rebound vigorously and create an intense shock. Two factors cooperate to achieve this result in the simulations. The first factor is the use of general relativity rather than the force field of Newtonian gravitation. The second is the assumption that nuclear matter is much more compressible than had been thought.

Baron's first result showed that a star of 12 solar masses would explode if the compressibility of nuclear matter is 1.5 times the standard value. This seemed rather arbitrary, but then one of us (Brown) examined the problem with a sophisticated method of nuclear-matter theory. It turned out that the most consistent interpretation of the experimental findings yields a compressibility of 2.5 times the standard value! We then found that in 1982 Andrew D. Jackson, E. Krotscheck, D. E. Meltzer, and R. A. Smith had reached the same conclusion by another method, but no one had recognized the relevance of their work to the supernova problem. We consider the higher estimate of nuclear compressibility quite reliable.

The mechanism described by Baron, Cooperstein, and Kahana seems to work for stars of up to about 18 solar masses. With still larger stars, however, even the powerful shock wave created in their simulations becomes stalled in the minefield. A star of 25 solar masses has about 2 solar masses of iron in its core, and so the shock wave must penetrate 1.2 solar masses of iron rather than .55 solar mass. The shock does not have enough energy to dissociate this much iron.

A plausible explanation of what might happen in these massive stars has recently emerged from the work of James R. Wilson of Lawrence Livermore, who has done extensive numerical simulations of supernova explosions. For some time it had seemed that when the shock wave failed, all the mass of the star might fall back into the core, which would evolve into a black hole. That fate is still a possible one, but Wilson noted a new phenomenon when he continued some of his simulations for a longer period.

In the collapsing stellar core it takes only 10 milliseconds or so for the shock wave to reach the minefield and stall. A simulation of the same events, even with the fastest computers, takes at least an hour. Wilson allowed his calculations to run roughly 100 times longer, to simulate a full second of time in the supernova. In almost all cases he found that the shock wave eventually revived.

The revival is due to heating by neutrinos. The inner core is a copious emitter of neutrinos because of continuing electron capture as the matter is compressed to nuclear density. Adam S. Burrows and Lattimer of Stony Brook and Mazurek have shown that half of the electrons in the homologous core are captured within about half a second, and the emitted neutrinos carry off about half of the gravitational energy set free by the collapse, some $10^{53}$ ergs. Deep within the core the neutrinos make frequent collisions with other particles; indeed, we noted above that they are trapped, in the sense that they cannot escape within the time needed for the homologous collapse. Eventually, though, the neutrinos do percolate upward and reach strata of lower density, where they can move freely.

At the radius where the shock wave stalls only one neutrino out of every 1,000 is likely to collide with a particle of matter, but these collisions nonetheless impart a significant amount of energy. Most of the energy goes into the dissociation of nuclei into nucleons, the very process that caused the shock to stall in the first place. Now, however, the neutrino energy heats the material and therefore raises the pressure sharply. We have named this period, when the shock wave stalls but is then revived by neutrino heating, "the pause that refreshes."

N eutrino heating is most effective at a radius of about 150 kilometers, where the probability of neutrino absorption is not too low and yet the temperature is not so high that the matter there is itself a significant emitter of neutrinos. The pressure increase at this radius is great enough, after about half a second, to stop the fall of the overlying matter and begin pushing it outward. Hence 150 kilometers becomes the bifurcation radius. All the matter within this boundary ultimately falls into the core; the matter outside, 20 solar masses or more, is expelled.

The one group of stars left to be considered are those of from eight to 11 solar masses, the smallest stars capable of supporting a type II supernova explosion. In 1980 Weaver and Woosley suggested that the stars in this group might form a separate class, in which the supernova mechanism is quite different from the mechanism in heavier stars.

According to calculations done by Nomoto, and by Weaver and Woosley, in the presupernova evolution of these lighter stars the core does not reach the temperature needed to form iron; instead fusion ends with a mixture of elements between oxygen and silicon. Energy production then stops, and since the mass of the core is greater than the Chandrasekhar limit, the core collapses. The shock wave generated by the collapse may be helped to propagate by two circumstances. First, breaking up oxygen or silicon nuclei robs the shock of less energy than the dissociation of iron nuclei would. Second, farther out in the star the density falls off abruptly (by a factor of roughly 10 billion) at the boundary between the carbon and the helium shells. The shock wave has a much easier time pushing through the lower-density material.

For a star of nine solar masses Nomoto finds that the presupernova core consists of oxygen, neon, and magnesium and has a mass of 1.35 solar masses. Nomoto and Wolfgang Hillebrandt of the Max-Planck Institute for Physics and Astrophysics in Munich have gone on to investigate the further development of this core. They find that the explosion proceeds easily through the core, aided by the burning of oxygen nuclei, and that a rather large amount of energy is released.

Two recent attempts to reproduce the Nomoto-Hillebrandt results have been unsuccessful, and so the status of their model remains unclear. We think the greater compressibility of nuclear matter assumed in the Baron-Cooperstein-Kahana program should be helpful here. Of course it is possible that stars this small do not give rise to supernovas; on the other hand, there are suggestive arguments (based on measurements of the abundance of various nuclear species) that the Crab Nebula was created by the explosion of a star of about nine solar masses.

After the outer layers of a star have been blown off, the fate of the core remains to be decided. Just as gravitation overwhelms electron pressure if the mass exceeds the Chandrasekhar limit, so even nuclear matter cannot resist compression if the gravitational field is strong enough. For a cold neutron star—one that has no source of supporting pressure other than the repulsion of nucleons—the limiting mass is thought to be about 1.8 solar masses. The compact remnant formed by the explosion of lighter stars is well below this limit, and so those supernovas presumably leave a stable neutron star. For the larger stars the question is in doubt. In Wilson's calculations any star of more than about 20 solar masses leaves a compact remnant of more than two solar masses. It would appear that the remnant will become a black hole, a region of space where matter has been crushed to infinite density.

Even if the compact remnant ultimately degenerates into a black hole, it begins as a hot neutron star. The central temperature immediately after the explosion is roughly 100 billion degrees Kelvin, which generates enough thermal pressure to support the star even if it is larger than 1.8 solar masses. The hot nuclear matter cools by the emission of neutrinos. The energy they carry off is more than 100 times the energy emitted in the explosion itself: $3 \times 10^{53}$ ergs. It is the energy equivalent of 10 percent of the mass of the neutron star.

The detection of neutrinos from a supernova explosion and from the subsequent cooling of the neutron star is one possible way we might get a better grasp of what goes on in these spectacular events. The neutrinos originate in the core of the star and pass almost unhindered through the outer layers, and so they carry evidence of conditions deep inside. Electromagnetic radiation, on the other hand, diffuses slowly through the shells of matter and reveals only what is happening at the surface. Neutrino detectors have recently been set up in mines and tunnels, where they are screened from the background of cosmic rays.

Another observational check on the validity of supernova models is the relative abundances of the chemical elements in the universe. Supernovas are probably the main source of all the elements heavier than carbon, and so the spectrum of elements released in simulated explosions ought to match the observed abundance ratios. Many attempts to reproduce the abundance ratios have failed, but earlier this year Weaver and Woosley completed calculations whose agreement with observation is surprisingly good. They began with Wilson's model for the explosion of a star of 25 solar masses. For almost all the elements and isotopes between carbon and iron their abundance ratios closely match the measured ones.

In recent years the study of supernovas has benefited from a close interaction between analytic theory and computer simulation. The first speculations about supernova mechanisms were put forward decades ago, but they could not be worked out in detail until the computers needed for numerical simulation became available. The results of the computations, on the other hand, cannot be understood except in the context of an analytic model. By continuing this collaboration we should be able to progress from a general grasp of principles and mechanisms to the detailed prediction of astronomical observations.

# Acknowledgments

Grateful acknowledgement is made to the following publications for permission to reprint previously published material.

HOW CLOSE IS THE DANGER? by Frederick Seitz and Hans A. Bethe was originally published in *One World or None*, edited by Dexter Masters and Katharine Way. London: Latimer House, Ltd., 1947.

THE HYDROGEN BOMB was adapted from an essay originally published in *Scientific American*, April 1950, pages 18–23. Copyright 1950 by Scientific American, Inc. All rights reserved.

BRIGHTER THAN A THOUSAND SUNS originally appeared as a book review of Robert Jungk's book of the same title in the *Bulletin of the Atomic Scientists*, December 1958, pages 426–428.

ULTIMATE CATASTROPHE? originally appeared in the *Bulletin of the Atomic Scientists*, June 1976, pages 36–37, as a rebuttal to an article by H. C. Dudley in the *Bulletin's* November 1975 issue.

THE CASE FOR ENDING NUCLEAR TESTS is reprinted from *The Atlantic Monthly*, August 1960, pages 43–51.

DISARMAMENT AND STRATEGY originally appeared in the *Bulletin of the Atomic Scientists*, September 1962, pages 14–22.

ANTIBALLISTIC-MISSILE SYSTEMS, by Richard L. Garwin and Hans A. Bethe, was adapted from an essay originally published in *Scientific American*, March 1968, pages 21–31. Copyright 1968 by Scientific American, Inc. All rights reserved.

MEANINGLESS SUPERIORITY originally appeared as an editorial in the *Bulletin of the Atomic Scientists*, October 1981, pages 1, 4.

WE ARE NOT INFERIOR TO THE SOVIETS appears in this collection for the first time. It was originally presented at a symposium of the American Physical Society's Forum on Physics and Society, April 1982.

THE FIVE-YEAR WAR PLAN, by Hans A. Bethe and Kurt Gottfried, is reprinted from *The New York Times*. Copyright 1982 by The New York Times Company. Reprinted with permission.

DEBATE: ELUSIVE SECURITY was originally published as two essays under the headline "Elusive Security: Do We Need More Nukes or Fewer?" in the *Washington Post*, February 6, 1983. The article by Hans A. Bethe and Kurt Gottfried was entitled "Save us all, Congress: no weapons in space, no unratified treaties." Senator Malcolm Wallop's contribution was entitled "Save American lives: build MX, two new missiles and space lasers."

SPACE-BASED BALLISTIC-MISSILE DEFENSE, by Hans A. Bethe, Richard L. Garwin, Kurt Gottfried and Henry W. Kendall, was adapted from an essay originally published in *Scientific American*, October 1984, pages 39–49. Copyright 1984 by Scientific American, Inc. All rights reserved.

THE TECHNOLOGICAL IMPERATIVE originally appeared in the *Bulletin of the Atomic Scientists*, August 1985, pages 34–36.

REDUCING THE RISK OF NUCLEAR WAR, by Hans A. Bethe and Robert S. McNamara, is reprinted from *The Atlantic Monthly*, July 1985, pages 43–51.

CHOP DOWN NUCLEAR ARSENALS was first published in the *Bulletin of the Atomic Scientists*, March 1989, pages 11–15. An earlier version of the article was presented as the Kistiakowsky Lecture to the American Academy of Arts and Sciences in Cambridge, Massachusetts on November 14, 1988.

THE VALUE OF A FREEZE, by Hans A. Bethe and Franklin A. Long, is reprinted from *The New York Times*, September 22, 1982. Copyright 1982 by The New York Times Company. Reprinted by permission.

DEBATE: BETHE VS. TELLER originally appeared on October 17, 1982 in *The Los Angeles Times* as two essays under the headline: "The Nuclear Freeze: Will it Really Make Us Safer?" Hans Bethe's article was entitled "Yes: It is a Significant Signal." Edward Teller's article was entitled "No: It is a Pointless Distraction."

AFTER THE FREEZE REFERENDUM originally appeared as an editorial in the *Bulletin of the Atomic Scientists* as "The Freeze Referendum: What Next?" by Hans A. Bethe and Franklin A. Long in February 1983, pages 2–3.

SCIENCE AND MORALITY originally appeared as an interview with Donald McDonald entitled "Scientist as Citizen" in *The Bulletin of the Atomic Scientists*, June 1962, pages 25–27.

BACK TO SCIENCE ADVISORS, by Hans A. Bethe and John Bardeen, is reprinted from *The New York Times*, May 17, 1986. Copyright 1986 by The New York Times Company. Reprinted by permission.

THE NECESSITY OF FISSION POWER was adapted from an essay originally published in *Scientific American*, January 1976, pages 21–31. Copyright 1976 by Scientific American, Inc. All rights reserved.

DEBATE: NUCLEAR SAFETY originally appeared as two articles in *The Bulletin of the Atomic Scientists*, September 1975, pages 37–41: "No Fundamental Change in the Situation" by Hans A. Bethe and "A Perspective on the Debate" by Frank von Hippel.

CHERNOBYL originally appeared as "U.S. Panel Assesses Chernobyl" in *The Bulletin of the Atomic Scientists*, December 1986, pages 45–46.

J. ROBERT OPPENHEIMER was originally published as "Oppenheimer: Where He Was There Was Always Life and Excitement" in *Science News and Comment*, Volume 155, March 3, 1967, pages 1080–1084. Copyright 1967 by the AAAS.

FREEMAN DYSON was originally published as a review of Dyson's book *Disturbing the Universe* in *Physics Today*, December 1979, pages 51–52.

HERMAN W. HOERLIN was originally published as an obituary by Hans A. Bethe, Donald M. Kerr, and Robert A. Jeffries in *Physics Today*, December 1984.

PAUL P. EWALD was originally published as an obituary by Helmut J. Juretschke, A. F. Moodie, H. K. Wagenfeld, and Hans A. Bethe in *Physics Today*, May 1986, pages 101–104.

RICHARD P. FEYNMAN was an obituary and was reprinted by permission from *Nature*, Volume 332, April 14, 1988, page 588. Copyright 1988 by Macmillan Magazines Limited.

ENERGY PRODUCTION IN STARS is adapted from the lecture delivered upon acceptance of the 1967 Nobel Prize for Physics. It also appeared in *Physics Today*, September 1968, pages 36–44. Copyright 1968 by the Nobel Foundation. The author is grateful to Professor E. E. Salpeter for his extensive help in preparing this essay.

HOW A SUPERNOVA EXPLODES, by Hans A. Bethe and Gerald Brown, was adapted from an essay originally published in *Scientific American*, May 1985, pages 60–68. Copyright 1985 by Scientific American, Inc. All rights reserved.

The author is also grateful to the following individuals for permission to reprint material authored or coauthored by them:

JOHN BARDEEN, professor emeritus of physics and electrical engineering at the University of Illinois, Urbana–Champaign.

GERALD BROWN, professor at the Institute for Theoretical Physics of the State University of New York at Stony Brook.

RICHARD L. GARWIN, IBM fellow and science advisor to the director of research at the IBM Research Division.

KURT GOTTFRIED, professor at the Newman Laboratory of Nuclear Studies of Cornell University.

ROBERT A. JEFFRIES, director of the Arms Control Technology Office of the Los Alamos National Laboratory.

HELMUT J. JURETSCHKE, professor of physics at Polytechnic University.

HENRY W. KENDALL, professor of physics at the Massachusetts Institute of Technology and chairman of the Union of Concerned Scientists.

DONALD M. KERR, executive vice president of EG&G Inc. and former director of the Los Alamos National Laboratory.

FRANKLIN A. LONG, professor emeritus of chemistry at Cornell University and a member of the Cornell University Peace Studies Program.

DONALD J. MCDONALD, former senior fellow, editor, and acting director of the Center for the Study of Democratic Institutions.

ROBERT S. MCNAMARA, former U.S. secretary of defense.

A. F. MOODIE, institute fellow in the Department of Applied Physics at the Royal Melbourne Institute of Technology.

FREDERICK SEITZ, president emeritus of Rockefeller University and past president of the National Academy of Sciences.

EDWARD TELLER, senior research fellow at the Hoover Institute at Stanford University and associate director emeritus at the Lawrence National Laboratory.

FRANK VON HIPPEL, faculty member of Princeton University's Woodrow Wilson School of Public and International Affairs and the university's Center for Energy and Environmental Studies.

MALCOLM WALLOP, U.S. senator from Wyoming.

H. K. WAGENFELD, head of the Department of Applied Physics at the Royal Melbourne Institute of Technology.

# Index